SpringerBriefs in Computer Science

For further volumes:
http://www.springer.com/series/10028

Dariusz Mrozek

High-Performance
Computational Solutions
in Protein Bioinformatics

 Springer

Dariusz Mrozek
Institute of Informatics, Silesian University
   of Technology
Gliwice
Poland

ISSN 2191-5768   ISSN 2191-5776 (electronic)
ISBN 978-3-319-06970-8   ISBN 978-3-319-06971-5 (eBook)
DOI 10.1007/978-3-319-06971-5
Springer Cham Heidelberg New York Dordrecht London

Library of Congress Control Number: 2014939535

Printed on acid-free paper

Springer is part of Springer Science+Business Media (www.springer.com)

For my beloved wife Bożena,
and my two sons Paweł and Henryk,
with all my love

Foreword by Jack Dongarra

High-performance computing most generally refers to the practice of aggregating computing power in a way that delivers much higher performance than one could get out of a typical desktop computer or workstation in order to solve large problems in science, engineering, or business. Big data is a popular term used to describe the exponential growth and availability of data, both structured and unstructured. The challenges include capture, curation, storage, search, sharing, transfer, analysis, and visualization. This timely book by Dariusz Mrozek gives you a quick introduction to the area of proteins and their structures, protein structure similarity searching carried out at main representation levels, and various techniques that can be used to accelerate similarity searches using high-performance computing and big data concepts. It presents introductory concepts of formal model of 3D protein structures for functional genomics, comparative bioinformatics, and molecular modeling, and the use of multithreading for efficient approximate searching on protein secondary structures. In addition, there is a material on finding 3D protein structure similarities accelerated with high-performance computing techniques.

The book is required reading to help in understanding for anyone working in the area of structural bioinformatics and biomedical databases and the use of high-performance computing. It explores the area of proteins and their structures in-depth and provides practical approaches to many problems that may be encountered. It is especially useful to applications developers, students, and teachers.

I have enjoyed and learned from this book and feel confident that you will as well.

Knoxville, April 2014   Jack Dongarra

Foreword by Albert Y. Zomaya

The field of Bioinformatics has undergone many advances over the last 20 years. Many of these advances are due to many developments in algorithmics and high-performance computing. The sub-field of proteomics is a major discipline in bioinformatics research and has great importance and this book deals with problems related to the structure of proteins. The book also shows how specialized computer architectures, such as GPUs and Cloud computing environments, can be used to accelerate the different computational problems.

I believe that the current book is a great addition to the existing literature on protein computations. It will serve as a source of up-to-date research in this continuously evolving area. The book also provides an opportunity for researchers to explore the use of advanced computer architectures and their impact on advancing our capabilities to conduct more sophisticated modeling and simulation studies.

The book should be well received by the research and development community and can be beneficial for graduate classes focusing on bioinformatics, computational biology, and systems biology.

Finally, I would like to congratulate Dr. Mrozek on a job well done, and I look forward to seeing the book in print.

Sydney, April 2014   Albert Y. Zomaya

Preface

For the last three decades we have been witnesses of the exponential growth of biological data in repositories such as GenBank, Protein Data Bank, UniProt/ SwissProt. The specificity of the data has inspired the scientific community to develop many algorithms that can be used to analyze the data and draw useful conclusions. A huge volume of the biological data caused many of the existing algorithms to become inefficient due to their computational complexity. Fortunately, the rapid development of computer science in the last decade has brought many technological innovations that can also be used in the field of bioinformatics and life sciences. The algorithms demonstrating a significant utility value, which have recently been perceived as too time-consuming, can now be efficiently used by applying the latest technological achievements like multithreading, Graphics Processing Units (GPUs), or cloud computing.

The book focuses on proteins and their structures, protein structure similarity searching carried out at main representation levels, and various techniques that can be used to accelerate similarity searches. The content of the book is divided into four parts. Part I provides a formal model of 3D protein structures for functional genomics, comparative bioinformatics, and molecular modeling. Part II focuses on the use of multithreading for efficient approximate searching on protein secondary structures. Parts III and IV concentrate on finding 3D protein structure similarities accelerated with the use of GPUs and cloud computing. Both parts describe the acceleration of different methods.

So, why proteins?, somebody can ask. I could answer the question by following Arthur M. Lesk in his book entitled *Introduction to Protein Science: Architecture, Function, and Genomics*. Because proteins are where the action is. But in fact, I have fallen in love with the beauty of protein structures at first sight inspired by the research conducted by R. I. P. Lech Znamirowski from the Silesian University of Technology in Gliwice, Poland. I decided to continue his research on proteins (Fig. 1).

Fig. 1 Preliminary architecture of the cloud-based solution for protein structure similarity searching drawn by me during the meeting (March 6th, 2013) with Artur Kłapciński, my associate in this project. Institute of Informatics, Silesian University of Technology, Gliwice, Poland

I believe this book will be interesting for scientists, researchers, and software developers working in the field of structural bioinformatics and biomedical databases. I hope that readers of the book will find it interesting and helpful in their everyday work.

Gliwice, April 2014   Dariusz Mrozek

Acknowledgments

Through many years there were many people involved in the research that I conducted. I find it hard to mention all of them. I would like to thank my wife Bożena Małysiak-Mrozek and also Bartek Socha, Miłosz Brożek, and Artur Kłapciński for their direct cooperation in my research leading to the emergence of the book. I would like to thank Alina Momot for her valuable advice on mathematical formulas and Henryk Małysiak for his mental support and constructive guidance resulting from the decades of experience in the academic and scientific work.

I would like to thank Microsoft Research for providing me a free access to computational resources of the Microsoft Azure cloud under the Microsoft Azure for Research Award program.

The emergence of this book was supported by the European Union from the European Social Fund (grant agreement number: UDA-POKL.04.01.01-00-106/09).

On a personal note, I would like to thank my family for all their love and patience in the moments of my absence resulting from my desire to write this book.

Contents

Acronyms

AFP	Aligned fragment pairs
BLOB	Binary large object
CUDA	Compute Unified Device Architecture
DBMS	Database management system
GPU	Graphics processing unit
IaaS	Infrastructure-as-a-Service
OODB	Object-oriented database
PaaS	Platform-as-a-Service
PDB	Protein Data Bank
RMSD	Root mean square deviation
SaaS	Software-as-a-Service
SIMD	Single instruction, multiple data
SIMT	Single instruction, multiple thread
SQL	Structured Query Language
SSE	Secondary structure element
SVD	Singular Value Decomposition
VM	Virtual machine
XML	Extensible Markup Language

Chapter 1
Formal Model of 3D Protein Structures for Functional Genomics, Comparative Bioinformatics, and Molecular Modeling

> *Proteins are where the action is.* Arthur M. Lesk, 2010
> *The great promise of structural bioinformatics is predicted on the belief that the availability of high-resolution structural information about biological systems will allow us to precisely reason about the function of these systems and the effects of modifications or perturbations.*
>
> Jenny Gu, Philip E. Bourne, 2009

Abstract This chapter shows how proteins can be represented in processes performed in scientific fields, such as functional genomics, comparative bioinformatics, and molecular modeling. The chapter begins with the general definition of protein spatial structure, which can be treated as a base for deriving other forms of representation. The general definition is then referenced to four representation levels of protein structure: primary, secondary, tertiary, and quaternary structure. This is followed by short description of protein geometry. And finally, at the end of the chapter, we will discuss energy features that can be calculated based on the general description of protein structure. The formal model defined in the chapter will be used in the description of algorithms presented in the following chapters of the book.

Keywords 3D protein structure · Formal model · Primary structure · Secondary structure · Tertiary structure · Quaternary structure · Energy features · Molecular modeling

1.1 Introduction

In the biological context, functioning of living organisms is tightly related to the presence and activity of proteins. Proteins are macromolecules that play a key role in all biochemical reactions in cells of living organisms. For this reason, they are said to be molecules of life. And indeed, they are involved in many processes, including:

D. Mrozek, *High-Performance Computational Solutions in Protein Bioinformatics*,
SpringerBriefs in Computer Science, DOI: 10.1007/978-3-319-06971-5_1,
© The Author(s) 2014

reaction catalysis (enzymes), energy storage, signal transmission, maintaining cell's cytoskeleton, immune response, stimuli response, cellular respiration, transport of small biomolecules, regulation of cell's growth and division.

Analyzing their general construction, proteins are macromolecules with the molecular mass above $10\,\text{kDa}$ ($1\,\text{Da} = 1.66 \times 10^{-24}\,\text{g}$) built up with amino acids (>100 amino acids, aa). Amino acids are linked to each other by peptide bonds forming a kind of linear chains [5]. Proteins can be described with the use of four representation levels: primary structure, secondary structure, tertiary structure, and quaternary structure. The last three levels define the protein conformation or protein spatial structure. The computer analysis of protein structures is usually carried out on one of the representation levels.

The computer analysis of protein spatial structure is very important from the viewpoint of the identification of protein functions, recognition of protein activity and analysis of reactions and interactions that the particular protein is involved in. This implies the exploration of various geometrical features of protein structures. There is no doubt that the structure of even small molecules are very complex, proteins are built up of hundreds of amino acids, and then thousands of atoms. This makes the computer analysis of protein structures more difficult and also influences a high computational complexity of algorithms for the analysis.

For any investigation related to protein bioinformatics it is essential to assume some representation of proteins as macromolecules. Methods that operate on proteins in scientific fields, such as functional genomics, comparative bioinformatics, and molecular modeling, usually assume a kind of model of protein structure. Formal models, in general, allow to define all concepts that are used in the area under consideration. They guarantee that all concepts that are used while designing and performing a process will be understood exactly as they are defined by an author of the method or procedure. This chapter attempts to capture the common model of protein structure which can be treated as a base model for the creation of dedicated models, derived either by the extension or the restriction, and used for the computations carried out in the selected area. In the following sections, we will discover a general definition of protein spatial structure and we will refer it to four representation levels of protein structure.

1.2 General Definition of Protein Spatial Structure

We define a 3D structure (S^{3D}) of protein P as a pair shown in Eq. (1.1).

$$S^{3D} = \langle A^{3D}, B^{3D} \rangle, \tag{1.1}$$

where A^{3D} is a set of atoms defined as follows:

$$A^{3D} = \left\{ a_n : n \in (1, \ldots, N) \ \wedge \ \exists f_E : A^{3D} \longrightarrow E \right\} \tag{1.2}$$

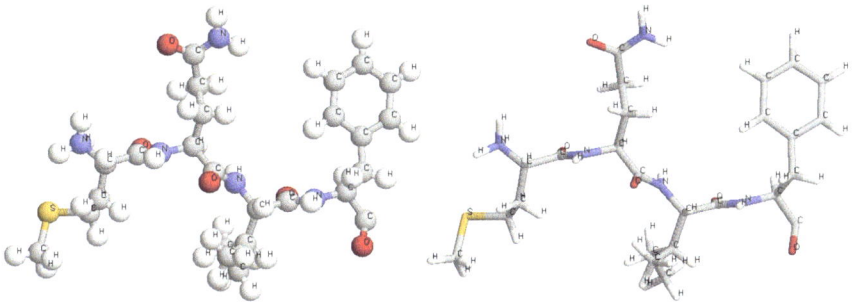

Fig. 1.1 Fragment of sample protein structure: (*left*) atoms and bonds, (*right*) bonds only. Colors and letters assigned to atoms distinguish their chemical elements. Visualized using RasMol [49]

where N is the number of atoms in a structure, f_E is a function which for each atom a_n assigns an element from the set of chemical elements E (e.g., N-nitrogen, O-oxygen, C-carbon, H-hydrogen, S-sulfur).

The B^{3D} is a set of bonds b_{ij} between two atoms $a_i, a_j \in A^{3D}$ defined as follows:

$$B^{3D} = \{b_{ij} : b_{ij} = (a_i, a_j) = (a_j, a_i) \ \wedge \ i, j \in (1, \dots, N)\}. \qquad (1.3)$$

Fragment of a sample protein structure is shown in Fig. 1.1.

Each atom a_n is described in three-dimensional space by Cartesian coordinates x, y, z:

$$a_n = (x_n, y_n, z_n)^T \quad \text{where} \quad x_n, y_n, z_n \in \mathbb{R}. \qquad (1.4)$$

Therefore, the length of bond b_{ij} between two atoms a_i and a_j can be calculated using the Pythagorean theorem:

$$\|b_{ij}\| = \sqrt{(x_i - x_j)^2 + (y_i - y_j)^2 + (z_i - z_j)^2}, \qquad (1.5)$$

which is equivalent to the norm calculation [8]:

$$\|b_{ij}\| = \|a_i - a_j\| = \sqrt{(a_i - a_j)^T (a_i - a_j)}. \qquad (1.6)$$

We can also state that:

$$a_n \in A^{3D} \implies \forall_{n \in \{1, \dots, N\}} \ \exists f_{Va} : A^{3D} \longrightarrow \mathbb{N}_+ \ \wedge \ \exists f_{Ve} : E \longrightarrow \mathbb{N}_+, \qquad (1.7)$$

where f_{Va} is a function determining the valence of an atom and f_{Ve} is a function determining the valence of chemical element. For example, $f_{Ve}(C) = 4$ and $f_{Ve}(O) = 2$.

1.3 A Reference to Representation Levels

Having defined such a general definition of protein spatial structure we can study what are the relationships between this structure and four main representation levels of protein structures, i.e., primary, secondary, tertiary, and quaternary structure. These relationships will be described in the following sections.

1.3.1 Primary Structure

Proteins are polypeptides built up with many, usually more than one hundred amino acids that are joined to each other by a peptide bond, and thus, forming a linear amino acid chain. The way how one amino acid joins to another, e.g., during the translation from the mRNA, is not accidental. Each amino acid has an N-terminus (also known as amino-terminus) and C-terminus (also known as carboxyl-terminus). When two amino acids join to each other, they form a peptide bond between C-terminus of the first amino acid and N-terminus of the second amino acid. When a single amino acid joins the forming chain during the protein synthesis, it links its N-terminus to the free C-terminus of the last amino acid in the chain. Therefore, the amino acid chain is created from N-terminus to C-terminus. Primary structure of protein is often represented as the amino acid sequence of the protein (also called protein sequence, polypeptide sequence), as it is presented in Fig. 1.2. The sequence is reported from N-terminus to C-terminus. Each letter in the sequence corresponds to one amino acid. Actually, the sequence is usually recorded in one-letter code, and rarely in three-letter code.

Protein sequence is determined by the nucleotide sequence of appropriate gene in the DNA. There are 20 standard amino acids encoded by the genetic code in the living organisms. However, in some organisms two additional amino acids can be encoded, i.e., selenocysteine and pyrrolysine. All amino acids differ in chemical properties and have various atomic constructions. Proteins can have one or many amino acid chains. The order of amino acids in the amino acid chain is unique and determines the function of the protein.

The representation of protein structure as a sequence of amino acids from Fig. 1.2a is very simple and frequently used by many algorithms and tools for protein comparison and similarity searching, such as Needleman-Wunsch [44] and Smith-Waterman [54] algorithms, BLAST [1] and FASTA [46] family of tools. The representation is also used by methods that predict protein structures from their sequences, like I-TASSER [61], Rosetta@home [29], Quark [63], and many others, e.g., [59] and [67].

Let us now reference the primary structure to the general definition of the spatial structure defined in the previous section. We can state that protein structure S^{3D} consists of M amino acids $P_m^{3D} \subsetneq S^{3D}$ such that:

$$P_m^{3D} = \langle A_m^P, B_m^P \rangle, \tag{1.8}$$

(a)
```
>2HBS| HOMO SAPIENS | DEOXYHEMOGLOBIN S
VLSPADKTNVKAAWGKVGAHAGEYGAEALERMFLSFPTTKTYFPHFDLSHGSAQVKGHGKKVADALTNAVA
HVDDMPNALSALSDLHAHKLRVDPVNFKLLSHCLLVTLAAHLPAEFTPAVHASLDKFLASVSTVLTSKYR
```
(b)
```
>2HBS| HOMO SAPIENS | DEOXYHEMOGLOBIN S
VAL LEU SER PRO ALA ASP LYS THR ASN VAL LYS ALA ALA TRP GLY LYS VAL GLY
ALA HIS ALA GLY GLU TYR GLY ALA GLU ALA LEU GLU ARG MET PHE LEU SER PHE
PRO THR THR LYS THR TYR PHE PRO HIS PHE ASP LEU SER HIS GLY SER ALA GLN
VAL LYS GLY HIS GLY LYS LYS VAL ALA ASP ALA LEU THR ASN ALA VAL ALA HIS
VAL ASP ASP MET PRO ASN ALA LEU SER ALA LEU SER ASP LEU HIS ALA HIS LYS
LEU ARG VAL ASP PRO VAL ASN PHE LYS LEU LEU SER HIS CYS LEU LEU VAL THR
LEU ALA ALA HIS LEU PRO ALA GLU PHE THR PRO ALA VAL HIS ALA SER LEU ASP
LYS PHE LEU ALA SER VAL SER THR VAL LEU THR SER LYS TYR ARG
```

Fig. 1.2 Primary structures of *Deoxyhemoglobin S* chain A in *Homo Sapiens* [PDB ID: 2HBS] [19]: **a** in a one-letter code describing amino acid types, **b** in a three-letter code describing amino acid types. First line provides some descriptive information

where

$$A_m^P \subsetneq A^{3D} \text{ and } B_m^P \subsetneq B^{3D}. \tag{1.9}$$

Sample protein P can be now recorded as a sequence of peptides p_m:

$$P = \left\{ p_m | i = 1, 2, \ldots, M \quad \wedge \quad \exists f_R : P \longrightarrow \Pi \right\}, \tag{1.10}$$

where M is a length of the sequence (in peptides), f_R is a function which for each peptide p_m assigns a type of amino acid from the set Π containing twenty (twenty two) standard amino acids.

Assuming that $p_m = P_m^{3D}$ we can associate the primary structure with the spatial structure S^{3D} (Fig. 1.3):

$$S^{3D} = \left\{ P_m^{3D} | m = 1, 2, \ldots, M \right\}. \tag{1.11}$$

Although

$$\bigcup_{m=1}^{M} P_m^{3D} \subsetneq S^{3D}, \tag{1.12}$$

in many situations related to processing of protein structures, we can assume that:

$$S^{3D} = \bigcup_{m=1}^{M} P_m^{3D}. \tag{1.13}$$

Fig. 1.3 Fragment of a sample protein structure showing the relationship between the primary structure and spatial structure. Successive amino acids are separated by *dashed lines*

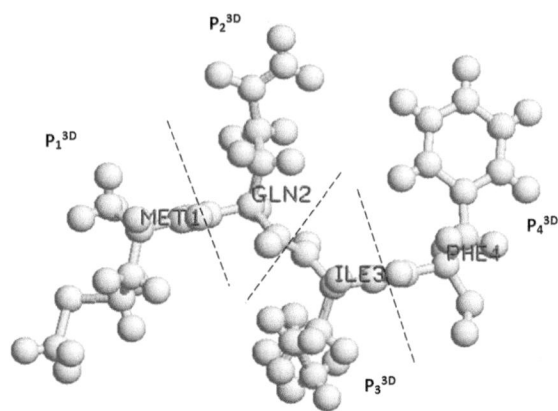

1.3.2 Secondary Structure

Secondary structure reveals specific spatial shapes in the construction of proteins. It shows how the linear chain of amino acids is formed in spiral α-helices, wavy β-strands, or loops. Indeed, these three shapes, α-helices, β-strands and loops, are main categories of secondary structures. Secondary structure itself does not describe the location of particular atoms in 3D space. It reflects local hydrogen interactions between some atoms of amino acids that are close in the amino acid chain.

Protein structure represented by means of secondary structure elements can have the following form:

$$S^S = \left\{ s_k^{se} | k = 1, 2, \ldots, K \quad \wedge \quad \exists f_S : S^S \longrightarrow \Sigma \right\}, \tag{1.14}$$

where s_k^{se} is the kth secondary structure element, K is the number of secondary structure elements in the protein, f_S is a function which for each element s_k^{se} assigns a type of secondary structure from the set Σ of possible secondary structure types. Actually, the f_S is a function that is sought by many researchers. Secondary structure prediction methods, like GOR [17], PREDATOR [15], or PredictProtein [48], try to model and implement the function in some way based on amino acid sequence.

In order to cover all parts of the protein structure the set Σ distinguishes four (sometimes more) types of secondary structures:

- α-helix,
- β-sheet or β-strand,
- loop, turn, or coil,
- and undetermined structure.

The first three types of secondary structures are visible in Fig. 1.7 (right) in the tertiary structure of a sample protein.

```
>Secondary structure for 2HBS chain A
U3 H32 U1 H6 U1 L3 U2 L3 U1 H19 L1 H3
L1 H9 L2 U3 L1 H17 U1 L3 U2 H19 L2 U2
```

Fig. 1.4 Secondary structures of *Deoxyhemoglobin S* chain A in *Homo Sapiens* [PDB ID: 2HBS]. First line provides some descriptive information

Each element s_k^{se} is characterized by two values:

$$s_k^{se} = [SSE_k, L_k], \tag{1.15}$$

where SSE_k describes the type of secondary structure (as mentioned above), $L_k \leq M$ is the length of the kth element s_k^{se} (measured in amino acids), M is a length of the amino acid chain. Such defined secondary structure can be represented as it is shown in Fig. 1.4, where particular symbols stand for: H—α-helix, E—β-strand, C/L—loop, turn or coil, U—unassigned structure.

The representation of protein secondary structures defined in Eqs. (1.14) and (1.15) and shown in Fig. 1.4 is used in some phases of the LOCK2 [52], CASSERT [39] and GPU-CASSERT [43] algorithms for 3D protein structure similarity searching, and in the indexing technique used in [18] and PSS-SQL [41] domain query languages.

Referencing the secondary structure to the general definition of the spatial structure we can state that a single element s_k^{se} is a substructure of the spatial structure S^{3D} containing usually several amino acids:

$$s_k^{se} = \langle A_k^S, B_k^S \rangle, \tag{1.16}$$

where

$$A_k^S \subsetneq A^{3D} \quad \text{and} \quad B_k^S = (B_k^{S*} \cup H_k) \subsetneq B^{3D}. \tag{1.17}$$

In formula (1.17) we take into account standard set of covalent bonds between atoms in the secondary structure s_k^{se}, represented by the B_k^{S*}, and additional hydrogen bonds stabilizing constructions of secondary structure elements, represented by the set H_k.

A spatial structure of sample protein can be now recorded as a sequence of secondary structure elements s_k^{se}:

$$S^{3D} = \left\{ s_k^{se} | k = 1, 2, \ldots, K \quad \wedge \quad \exists f_L : A_k^S \longrightarrow \mathbb{R}^3 \right\} \tag{1.18}$$

where K is the number of secondary structure elements in the protein, f_L is a function which for each atom a_n of the secondary structure s_k^{se} assigns a location in space described by Cartesian coordinates (x_n, y_n, z_n). There are many approaches to modeling the function f_L and finding the Cartesian coordinates for atoms of the protein structure. Physical methods rely on physical forces and interactions between atoms in a protein. Representatives of the approach include already mentioned

```
>Sequence and secondary structure for 2HBS chain A
1      VLSPADKTNV KAAWGKVGAH AGEYGAEALE RMFLSFPTTK TYFPHFDLSH
       UUUHHHHHHH HHHHHHHHHH HHHHHHHHHH HHHHHUHHHH HHULLLUULL
51     GSAQVKGHGK KVADALTNAV AHVDDMPNAL SALSDLHAHK LRVDPVNFKL
       LUHHHHHHHH HHHHHHHHHH HLHHHHHHHL HHHHHHHHHL LUUULHHHHH
101    LSHCLLVTLA AHLPAEFTPA VHASLDKFLA SVSTVLTSKY R
       HHHHHHHHHH HHULLLUUHH HHHHHHHHHH HHHHHHHLLU U
```

Fig. 1.5 Secondary structure and primary structure of *Deoxyhemoglobin S* chain A in *Homo Sapiens* [PDB ID: 2HBS]. First line provides some descriptive information

I-TASSER [61], Rosetta@home [29], Quark [63], WZ [59] and NPF [67]. Comparative methods rely on already known structures that are deposited in macromolecular data repositories, such as Protein Data Bank (PDB) [4]. Representatives of the comparative approaches are Robetta [26], Modeler [13], RaptorX [24], HHpred [56], Swiss-Model [2] for homology modeling, and Sparks-X [64], Raptor [62], and Phyre [25] for fold recognition.

It is also interesting to follow the relationship between protein secondary structure and primary structure. We can record a single element s_k^{se} as a subsequence of amino acids:

$$s_k^{se} = (p_l, p_{l+1}, \ldots, p_m), \quad \text{where} \quad 1 \leq l \leq m \leq M, \tag{1.19}$$

and where element p is any amino acid forming part of the secondary structure s_k^{se}, and M is a length of the protein (in amino acids).

It can be also noted that for any $p_m = P_m^{3D} = \langle A_m^P, B_m^P \rangle$:

$$A_m^P \subseteq A_k^S \quad \text{and} \quad B_m^P \subseteq B_k^S. \tag{1.20}$$

Such a relationship between secondary structure and primary structure is usually represented as it is shown in Fig. 1.5 and can be visualized in a similar fashion to that shown in Fig. 1.6. The representation of protein secondary structures shown in Fig. 1.5 is used as one of the protein geometry features in algorithms for 3D protein structure similarity searching, e.g., CTSS [9] and mentioned CASSERT [39].

1.3.3 Tertiary Structure

Tertiary structure is a higher degree of organization. Proteins achieve their tertiary structures through the protein folding process. In this process a polypeptide chain acquires its correct three-dimensional structure and adopts biologically active native state [5]. Many proteins have only one amino acid chain, so that tertiary structure is enough to describe their spatial structure. For those that are composed of more than one chain, the quaternary structure is required.

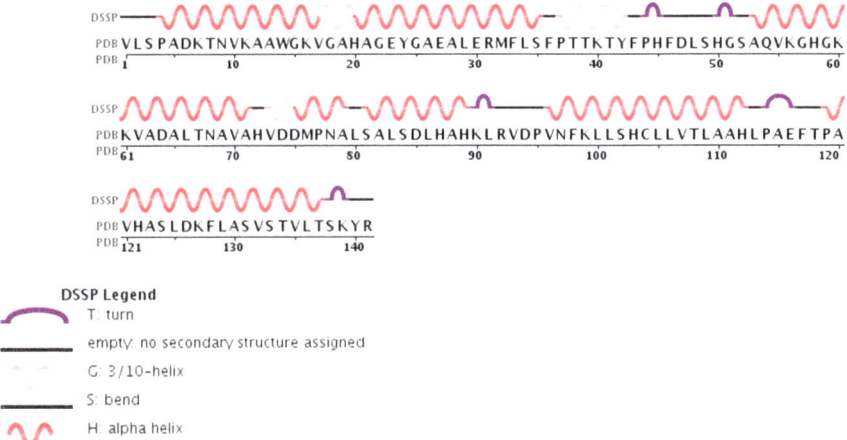

Fig. 1.6 Relationship between secondary structure and primary structure of *Deoxyhemoglobin S* chain A in *Homo Sapiens* [PDB ID: 2HBS] visualized graphically at the Protein Data Bank [4] web site (http://www.pdb.org, accessed on March 7th, 2014)

Tertiary structure requires 3D coordinates of all atoms of the protein structure to be determined. Therefore, we can state that if the number of polypeptide chains $H = 1$ the general spatial structure S^{3D} describes the tertiary structure S^{T} of a protein:

$$H = 1 \iff S^{T} = S^{3D}, \tag{1.21}$$

and:

$$S^{T} = \langle A^{T}, B^{T} \rangle. \tag{1.22}$$

At this point, description of tertiary structure is the same as the description of the general spatial structure S^{3D}. Example of tertiary structure is presented in Fig. 1.7.

From the viewpoint of secondary structures, the tertiary structure specifies positional relationships of secondary structures [8], which is presented in Fig. 1.7 (right). The set of atoms of the tertiary structure A^{T} consists of atoms forming all of the secondary structures packed into the protein structure (represented as the set A^{T*}). It also includes possible atoms from additional functional groups (represented as the set A^{FG}), e.g., prosthetic groups, inhibitors, solvent molecules, and ions for which coordinates are supplied. Example of prosthetic group is shown in Fig. 1.8. Similarly, in addition to covalent and noncovalent bonds between atoms forming amino acids of the protein chain (represented as the set B^{T*}):

$$B^{T*} = \bigcup_{m=1}^{M} B_{m}^{P}, \tag{1.23}$$

Fig. 1.7 Tertiary structure
of sample protein *Cyclin
Dependent Kinase CDK2*
[PDB ID: 1B38] [7]: (*left*)
representation showing atoms
and bonds, (*right*) representation showing secondary
structures and their relative
orientation. Visualized using
RasMol [49]

Fig. 1.8 Prosthetic heme
group responsible for oxygen
binding, distinguished in the
structure of *Myoglobin* [PDB
ID: 1MBN] [60]

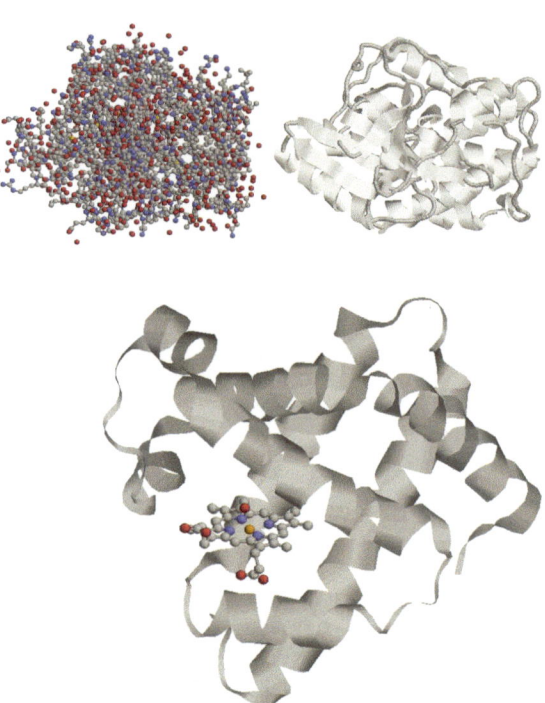

the set of bonds of the tertiary structure B^T may also consist of bonds between
atoms from the functional groups (represented as the set B^{FG}) and additional bonds
stabilizing the tertiary structure (represented as the set B^{stab}), e.g., disulfide bridges
(S–S) between cysteine residues (Fig. 1.9). Therefore:

$$A^T = A^{T*} \cup A^{FG} \quad \wedge \quad B^T = B^{T*} \cup B^{FG} \cup B^{stab}. \tag{1.24}$$

The representation of the 3D protein structure, having regard to formulas (1.21)–
(1.24) and earlier formulas (1.1)–(1.7), is used by many algorithms for protein structure alignment and similarity searching, including DALI [21], LOCK2 [52], FATCAT
[65], CTSS [9], CE [53], FAST [66], and others [39]. To complete the search task,
these algorithms usually does not explore whole sets of atoms A^T and bonds B^T, but
use reduced sets $A^{T'}$ of chosen atoms, e.g., C_α atoms of the backbone, and distances
between the atoms (calculated using the formula (1.5) or (1.6)):

$$A_\alpha^{T'} = \left\{ a_m | m = 1, 2, \ldots, M \quad \wedge \quad \forall m \ a_m \in A_m^P \wedge f_{Va}(a_m) = C_\alpha \right\}, \tag{1.25}$$

where M is the length of protein chain (in residues).

Fig. 1.9 Disulfide bridge
between two sulfur atoms
in cysteine residues in
sample protein *Glutaredoxin
-1-Ribonucleotide Reductase
B1* [PDB ID: 1QFN] [3]

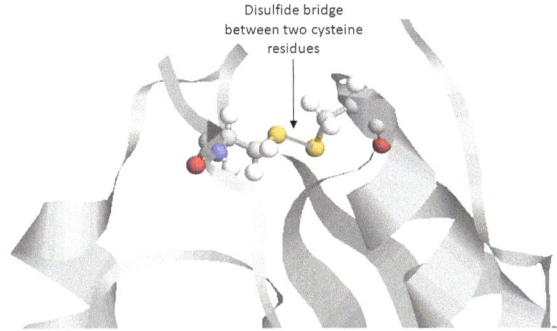

Disulfide bridge
between two cysteine
residues

Some algorithms, like SSAP [58], also use the C_β atoms in order to include an information on the orientation of the side chains:

$$A_\beta^{T'} = \{a_m | m = 1, 2, \dots, M' \ \wedge \ M' \leq M$$
$$\wedge \ \forall m \ a_m \in A_m^P \wedge f_{Va}(a_m) = C_\beta\}, \tag{1.26}$$

or combinations of the two atoms:

$$A_{\alpha\beta}^{T'} \subset A_\alpha^{T'} \cup A_\beta^{T'}. \tag{1.27}$$

Molecular viewers, like Chime [11], QMOL [16], Jmol [23], PMV [40], RasMol [49], PyMOL [50], and MViewer [57], also make use of the whole set of atoms A^T and bonds B^T or just subsets of them (depending of the display mode) during protein structure visualization. For example, in the *balls and sticks* display mode (Fig. 1.7, left) they use whole set of atoms and bonds, and in the *backbone* mode they use just positions of the C_α atoms to display the protein backbone.

1.3.4 Quaternary Structure

Quaternary structure describes spatial structures of proteins that have more than one polypeptide chain. Quaternary structure shows mutual location of tertiary structures of these chains in the three-dimensional space. Therefore, we can represent a quaternary structure as follows:

$$S^Q = \{c_h | h = 1, 2, \dots, H$$
$$\wedge \ \exists f_{CID} : S^Q \longrightarrow \{A, B, C, \dots, X, Y, Z\}$$
$$\wedge \ \exists f_T : S^Q \longrightarrow S^T\}. \tag{1.28}$$

Fig. 1.10 Quaternary structure of *Human Deoxyhemoglobin* [PDB ID: 4HHB] [14] containing four chains and heme

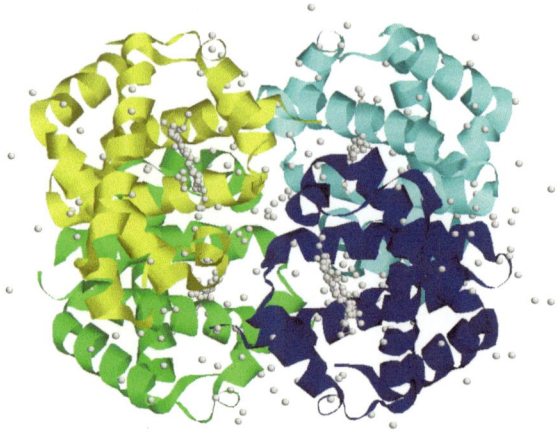

where H is the number of protein chains, f_{CID} is a function which for each chain c_h of the quaternary structure S^Q assigns a chain identifier, e.g., A, B, ..., Z, and f_T is a function which for each chain c_h of the quaternary structure S^Q assigns tertiary structure S^T.

Therefore, we can state that if the number of polypeptide chains $H > 1$ the general spatial structure S^{3D} describes the quaternary structure S^Q of a protein:

$$H > 1 \Longleftrightarrow S^Q = S^{3D}. \tag{1.29}$$

Such protein structures that are composed of a number of chains are called oligomeric complexes [8]. Examples of quaternary structures are shown in Figs. 1.10 and 1.11.

If each chain c_h has its tertiary structure, we can note that:

$$c_h = \langle A_h^T, B_h^T \rangle, \tag{1.30}$$

and that:

$$A^{3D} = A^Q = \left(\bigcup_{h=1}^{H} A_h^T \right) \cup A^{FG}, \tag{1.31}$$

$$B^{3D} = B^Q = \left(\bigcup_{h=1}^{H} B_h^T \right) \cup B^{FG} \cup B^{stab}. \tag{1.32}$$

Again, the set of atoms A^Q forming quaternary structure of a protein consists of atoms belonging directly to particular component polypeptide chains (A_h^T) and atoms of additional functional groups (A^{FG}). The set of bonds B^Q consists of covalent

Fig. 1.11 Quaternary structure of *Insulin Hormone* [PDB ID: 1ZNJ] [55] containing six chains and zinc atoms

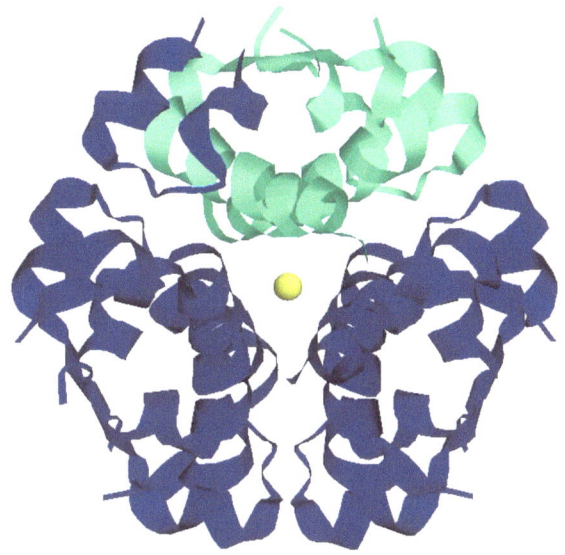

bonds linking atoms of each of the polypeptide chains B_h^T, bonds linking atoms of functional groups B^{FG}, and bonds stabilizing the quaternary structure B^{stab}, e.g., intra-chain disulfide bridges.

1.4 Relative Coordinates of Protein Structures

Some of the computational processes prefer to use relative coordinates, rather than absolute coordinates of particular atoms of protein structures. For example, in protein structure prediction by energy minimization many different relative coordinates are used while performing a computational process. These relative coordinates can be derived based on the protein structure S^{3D}, for which absolute coordinates are being known.

We have already had the opportunity to see one of the relative coordinates when we talked about a set of bonds, the B^{3D} component of the protein structure S^{3D}. These were **bond lengths**. Bond lengths (Fig. 1.12) were studied intensively during past years and after making some statistics we know that lengths of bonds between particular types of atoms in protein backbone are similar. Bond length for $N - C_\alpha$ is 1.47 Å(1 Å $= 10^{-10}$ m), for $C_\alpha - C$ is 1.53 Å, and for $C - N$ is 1.32 Å [51]. However, investigation of differences and similarities between bond lengths is still interesting. Some computational procedures require bond lengths to be calculated. For example, while comparing two protein structures selected types of bonds, like $C_\alpha - C'$, can be compared for each pair of compared amino acids. Bond lengths are also used while calculating bond stretching component energy of total potential energy of protein

Fig. 1.12 Graphical interpretation of bond length (*top
left*), interatomic distance
(*top right*), and bond angle
(*bottom*)

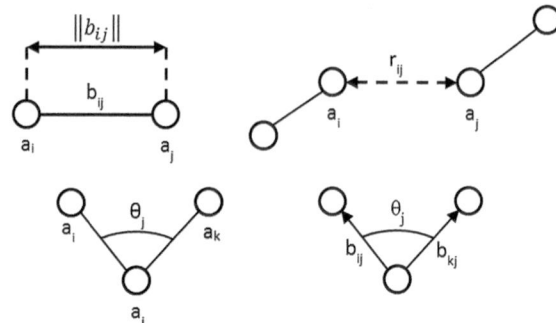

structure (Sect. 1.5). Bond lengths can be calculated according to formulas (1.5) and
(1.6) shown earlier in this chapter.

A kind of generalization of bond lengths can be **interatomic distances**. Interatomic distances describe the distance between two atoms (Fig. 1.12). However,
these atoms do not have to be connected by any bond. Interatomic distances can
be calculated according to the same formulas (1.5) and (1.6) as bond lengths. And
they are very useful when we want to study interactions between particular atoms in
protein structure or between atoms of two molecules, e.g., two substrates of cellular
reaction. They are also frequently calculated in protein structure comparison. For
example, popular DALI algorithm [21] uses distances between C_α atoms in order
to calculate the so-called *distance matrices* that represent protein structures in the
comparison process.

Another relative feature, which is studied by researchers in the field of chemistry
and molecular biology, is **bond angles**. Bond angles or **valence angles** are, next to
the bond lengths, the principal relative features that control the shape of 3D protein
structures. In order to calculate a bond angle, we have to know the positions of three
atoms (Fig. 1.12).

The angle between two bonds b_{ij} and b_{kj} linking these three atoms can be calculated from a dot product of their respective vectors (Fig. 1.12, bottom right):

$$\cos \theta_j = \frac{b_{ij} \cdot b_{kj}}{\|b_{ij}\| \|b_{kj}\|} \tag{1.33}$$

A very important information for the analysis of 3D protein structures bring also
torsion angles. Torsion angles are dihedral angles that describe the rotation of protein
polypeptide backbone around particular bonds. There are three types of torsion angles
that are calculated for protein structures, i.e., Phi (ϕ), Psi (ψ), and Omega (ω). The
Phi torsion angle describes the rotation around the $N - C_\alpha$ bond, the Psi torsion angle
describes the rotation around the $C_\alpha - C'$ bond, and Omega torsion angle describes
the rotation around the $C' - N$ bond (see Fig. 1.13).

Looking at Fig. 1.13 we can notice that the peptide bond formed by C, O, N,
H atoms is planar which restricts the rotation around the $C' - N$ bond. The Omega

Fig. 1.13 Overview of
protein construction. Visible atoms forming the main
chain of the polypeptide. Side
chains are marked as R_0, R_1,
R_2. Reproduced from [68]

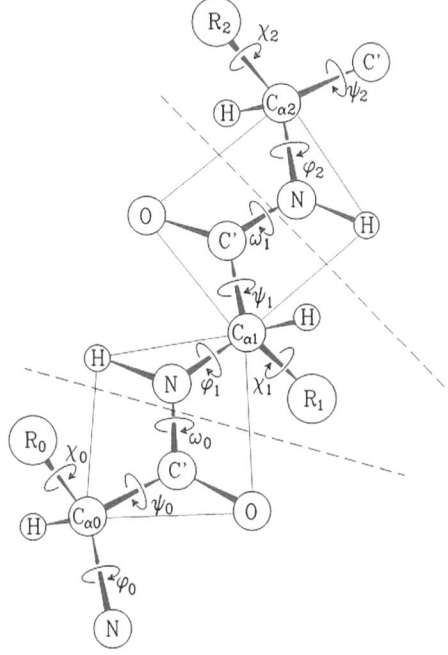

angle is then essentially fixed to 180° due to the partial double-bond character of the
peptide bond. Therefore, main chain rotations are restricted to the Phi and Psi angles,
and these angles provide the flexibility required for folding the protein backbone.
This information is utilized in algorithms for protein structure prediction, e.g., WZ
[59] and NPF [67], that model protein structure by random choosing and rotating the
Phi and Psi angles.

Torsion angles can be calculated using the dot product of the normal vectors of
the two planes defined by three successive atoms a_i, a_j, a_k and a_j, a_k, a_l as shown
in Fig. 1.14.

These normals can be calculated from the cross products of vectors:

$$n_1 = b_{ij} \times b_{kj} \quad \text{and} \quad n_2 = b_{jk} \times b_{lk}, \tag{1.34}$$

creating particular planes (vectors defined by three successive atoms a_i, a_j, a_k and
a_j, a_k, a_l) and then, used to calculate a dihedral angle from the dot product $n_1 \cdot n_2$:

$$\cos \omega = \frac{n_1 \cdot n_2}{\|n_1\| \|n_2\|}, \tag{1.35}$$

where ω is a calculated torsion angle (Phi, Psi, Omega), depending on which successive atoms of the backbone are inserted in place of a_i, a_j, a_k, a_l. For the Phi torsion
angle these should be atoms $C'_{i-1} - N_i - C\alpha_i - C'_i$, for the Psi these should be
$N_i - C\alpha_i - C'_i - N_{i+1}$, and for Omega these should be $C\alpha_i - C'_i - N_{i+1} - C\alpha_{i+1}$.

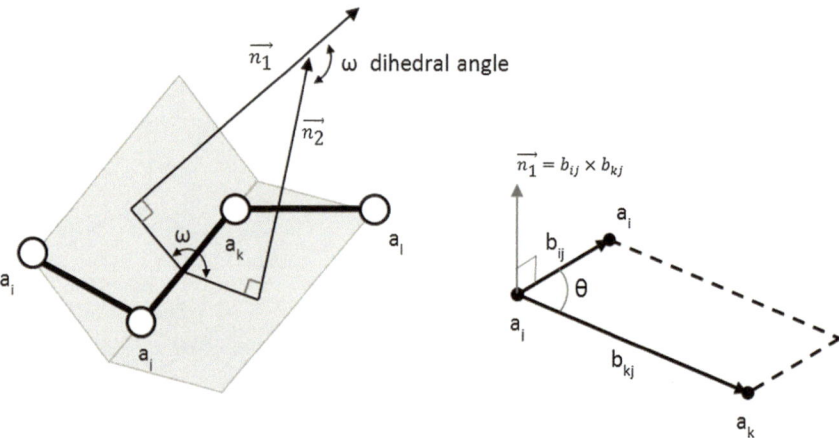

Fig. 1.14 Calculation of the dihedral angle between two planes based on normal vectors (*left*). Calculation of the normal vector as a cross product of vectors defining a plane (*right*). Redrawn based on [8]

Fig. 1.15 Ramachandran plot showing distribution of torsion angles Phi and Psi for a sample protein structure. Generated by PROCHECK program [27]

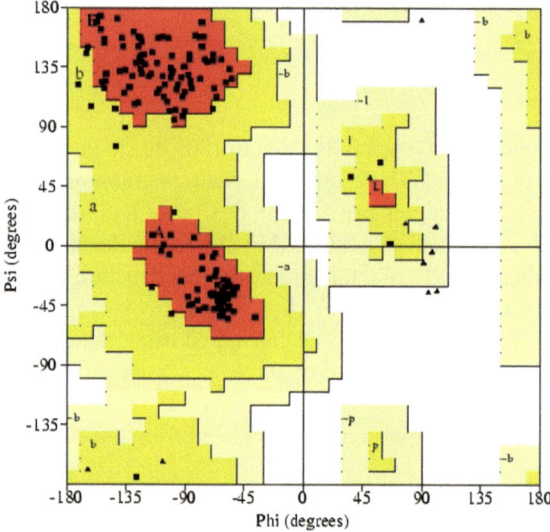

Theoretically, the Phi and Psi angles can take values ranging from −180° to 180°. However, in protein molecules rotations of Phi and Psi torsion angles are restricted to certain values due to sterical collisions between main chain and side chain atoms. Moreover, protein regions that form a particular secondary structure, impose additional constraints on the values of these torsion angles. This was noticed by Ramachandran and colleagues in [47]. The chart showing real-values of the Phi and Psi angles and possible combinations of these values for various types of secondary structures is known today as **Ramachandran plot** (Fig. 1.15).

On the Ramachandran plot, values of the Phi angle are plotted on the x-axis and values of the Psi angle are plotted on the y-axis. For many years the Ramachandran plot has been widely used by scientists in order to validate torsion angles and asses the quality and correctness of protein structures that were obtained by means of experimental methods (X-ray crystallography and NMR spectroscopy) or by homology modeling [13]. For example, Ramachandran plots are created by the popular PROCHECK, a program that provides a detailed check on the stereochemistry of a protein structures [27].

1.5 Energy Properties of Protein Structures

Protein structure S^{3D} can be also analyzed in terms of forces that act on an each atom within the molecule. In such an approach, atoms are considered as masses that interact with each other. Various forces between interacting atoms cause changes in the potential energy of the molecular system S^{3D}. The molecular system can be then modeled by molecular mechanics, where the potential energy of the set of atoms A^{3D} is described by empirical **force fields** providing a functional form for the potential energy and containing a set of parameters for particular atoms in the set A^{3D}. This kind of description of the molecular system S^{3D} usually takes place while studying molecular dynamics of proteins or modeling the protein structure by minimizing the conformational energy. Scientists assume here that when a protein stabilizes the positions of its atoms, the energy of such a molecular system is minimized. Consequently, any changes in the protein conformation causing deviations of bond lengths, angles, and intermolecular distances from reference values come with energy sanctions [28].

There are various types of force fields that were derived experimentally or by using quantum mechanical calculations. The most popular ones include: AMBER (Assisted Model Building and Energy Refinement) [12], CHARMM (Chemistry at HARvard Molecular Mechanics) [6], and GROMOS (GROningen MOlecular Simulation package) [45], but there are also many others. These force fields provide different functional forms that model the potential energy of the molecular system S^{3D}. However, they usually contain the following common energy terms:

$$E_T(S^{3D}) = E_{BS} + E_{AB} + E_{TA} + E_{VDW} + E_{CC}, \qquad (1.36)$$

where $E_T(S^{3D})$ denotes the total potential energy, and particular component energies contributing to the total potential E_T are defined as follows:

- bond stretching (E_{BS})

$$E_{BS}(S^{3D}) = \sum_{j=1}^{bonds} \frac{k_j}{2} \left(d_j - d_j^0 \right)^2, \qquad (1.37)$$

where k_j is a bond stretching force constant, d_j is a distance between two atoms (real bond length), d_j^0 is an optimal bond length;

- angle bending (E_{AB})

$$E_{AB}(S^{3D}) = \sum_{j=1}^{\text{angles}} \frac{k_j}{2} \left(\theta_j - \theta_j^0\right)^2,$$ (1.38)

where k_j is a bending force constant, θ_j is an actual value of the valence angle, θ_j^0 is an optimal valence angle;

- torsional angle (E_{TA})

$$E_{TA}(S^{3D}) = \sum_{j=1}^{\text{torsions}} \frac{V_j}{2}(1 + \cos(n\omega - \gamma)),$$ (1.39)

where V_j denotes the height of the torsional barrier, n is a periodicity, ω is the torsion angle, γ is a phase factor;

- van der Waals (E_{VDW})

$$E_{VDW}(S^{3D}) = \sum_{k=1}^{N} \sum_{j=k+1}^{N} \left(4\varepsilon_{kj}\left[\left(\frac{\sigma_{kj}}{r_{kj}}\right)^{12} - \left(\frac{\sigma_{kj}}{r_{kj}}\right)^{6}\right]\right),$$ (1.40)

where r_{kj} denotes the distance between atoms k and j, σ_{kj} is a collision diameter, ε_{kj} is a well depth, and N is the number of atoms in the structure S^{3D};

- electrostatic (E_{CC}), also known as Coulomb or charge-charge

$$E_{CC}(S^{3D}) = \sum_{k=1}^{N} \sum_{j=k+1}^{N} \frac{q_k q_j}{4\pi \varepsilon_0 r_{kj}},$$ (1.41)

where q_k, q_j are atomic charges, r_{kj} denotes the distance between atoms k and j, ε_0 is a dielectric constant, and N is the number of atoms in the structure S^{3D}.

The first three terms are called as *bonded interactions*, since they occur between atoms that are covalently bonded. Their graphical interpretation is shown in Fig. 1.16. The last two terms are referred as *nonbonded interactions*, since they occur between nonbonded atoms. Graphical interpretations of these two terms are shown in Fig. 1.17.

There can be more energy terms in the function describing the total potential energy. Further description of these and other component energies is out of the scope of the book. However, readers who are interested in details of these potentials are encouraged to read the book of the Leach [28].

As can be also seen, calculations of the potential energy require both components A^{3D} and B^{3D} of the defined model of the protein structure S^{3D}, as well as some of the relative coordinates that can be derived from the structure. It is also worth noting

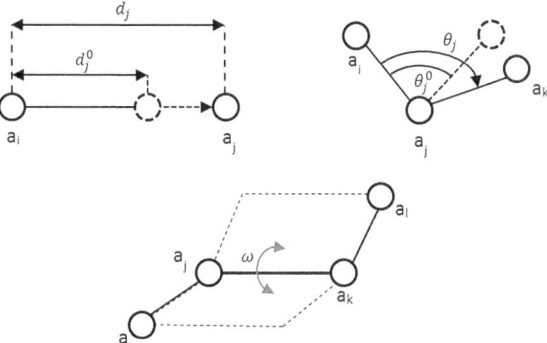

Fig. 1.16 Schematic interpretation of bonded interactions: (*top left*) bond stretching, (*top right*) angle bending, and (*bottom*) torsional angle

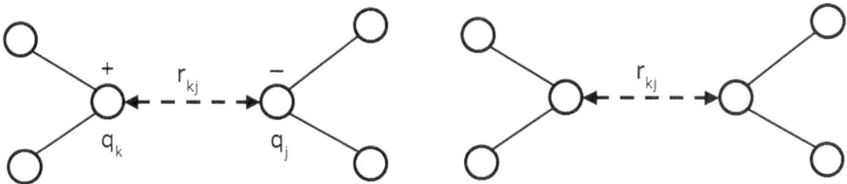

Fig. 1.17 Schematic interpretation of nonbonded interactions: (*left*) electrostatic and (*right*) van der Waals

methods that make use of energy properties of protein structures for investigating protein sequence-structure-function relationships, protein conformational modifications [33], and protein activity in cellular reactions [10, 22, 34] through energy properties. Representatives of the methods are ePros [20], and successive versions of the EAST method [30–32], including FN-EAST [37] and FS-EAST [38, 42], that use the EDB database [36] and EDML data exchange format [35].

1.6 Summary

The model of protein structure S^{3D} shown in this chapter has a general purpose and can be used while describing protein molecules in many different processes related to functional genomics, comparative biology, and molecular modeling. In fact, protein structures can be described by many various features and those presented in this chapter do not cover all of them. Which features are used, depend on the particular process. However, most of them, if not all, can be derived from the general model S^{3D}.

The general model of protein structure shown in this chapter is especially useful in any process related to protein modeling, drug design, or protein structure comparison. In these processes acting at the level of individual atoms and inspection of their positions is particularly important. Some of the methods and particular representations of protein structures will be shown in the following chapters.

References

1. Altschul, S.F., Gish, W., Miller, W., Myers, E.W., Lipman, D.J.: Basic local alignment search tool. J. Mol. Biol. **215**, 403–410 (1990)
2. Arnold, K., Bordoli, L., Kopp, J., Schwede, T.: The SWISS-MODEL workspace: a web-based environment for protein structure homology modelling. Bioinformatics **22**(2), 195–201 (2009)
3. Berardi, M., Bushweller, J.: Binding specificity and mechanistic insight into glutaredoxin-catalyzed protein disulfide reduction. J. Mol. Biol. **292**, 151–161 (1999)
4. Berman, H., et al.: The Protein Data Bank. Nucleic Acids Res. **28**, 235–242 (2000)
5. Branden, C., Tooze, J.: Introduction to Protein Structure, 2nd edn. Garland Science, New York (1999)
6. Brooks, B.R., Bruccoleri, R.E., Olafson, B.D., States, D.J., Swaminathan, S., Karplus, M.: CHARMM: a program for macromolecular energy, minimization, and dynamics calculations. J. Comp. Chem. **4**(2), 187–217 (1983)
7. Brown, N., Noble, M., Lawrie, A., Morris, M., et al.: Effects of phosphorylation of threonine 160 on cyclin-dependent kinase 2 structure and activity. J. Biol. Chem. **274**, 8746–8756 (1999)
8. Burkowski. F.: Structural Bioinformatics: An Algorithmic Approach, 1st edn. Chapman and Hall/CRC, Boca Raton (2008)
9. Can, T., Wang, Y.: CTSS: a robust and efficient method for protein structure alignment based on local geometrical and biological features. In: Proceedings of the 2003 IEEE Bioinformatics Conference (CSB 2003), pp. 169–179 (2003)
10. Chen, P.Y., Lin, K.C., Lin, J.P., et al.: Phenethyl isothiocyanate (PEITC) inhibits the growth of human oral squamous carcinoma HSC-3 cells through G0/G1 phase arrest and mitochondria-mediated apoptotic cell death. Evidence-Based Complementary and Alternative Medicine, vol. 2012. Article ID 718320, pp. 1–12 (2012)
11. Chime and Jmol Homepage: Molecular Visualization Resources. http://www.umass.edu/microbio/chime/
12. Cornell, W.D., Cieplak, P., Bayly, C.I., Gould, I.R., Merz, K.M. Jr., Ferguson, D.M., Spellmeyer, D.C., Fox, T., Caldwell, J.W., Kollman, P.A.: A second generation force field for the simulation of proteins, nucleic acids, and organic molecules. J. Am. Chem. Soc. **117**, 5179–5197 (1995)
13. Eswar, N., Marti-Renom, M.A., Webb, B., Madhusudhan, M.S., Eramian, D., Shen, M., Pieper, U., Sali, A.: Comparative Protein Structure Modeling with MODELLER. Current Protocols in Bioinformatics, Supplement 15, pp. 5.6.1–5.6.30. John Wiley & Sons Inc, New York (2006)
14. Fermi, G., Perutz, M.F., Shaanan, B., Fourme, R.: The crystal structure of human deoxy-haemoglobin at 1.74 A resolution. J. Mol. Biol. **175**, 159–174 (1984)
15. Frishman, D., Argos, P.: 75% accuracy in protein secondary structure prediction. Proteins **27**, 329–335 (1997)
16. Gans, J., Shalloway, D.: Qmol: a program for molecular visualization on windows-based PCs. J. Mol. Graph Model **19**(6), 557–559 (2001)
17. Garnier, J., Gibrat, J.F., Robson, B.: GOR method for predicting protein secondary structure from amino acid sequence. Methods Enzymol. **266**, 540–553 (1996)
18. Hammel, L., Patel, J.M.: Searching on the secondary structure of protein sequences. In: Proceedings of the 28th International Conference on Very Large Data Bases. Hong Kong, China, pp. 634–645 (2002)
19. Harrington, D., Adachi, K., Royer Jr, W.: The high resolution crystal structure of deoxyhemoglobin S. J. Mol. Biol. **272**, 398–407 (1997)

20. Heinke, F., Schildbach, S., Stockmann, D., Labudde, D.: eProSa database and toolbox for investigating protein sequence-structure-function relationships through energy profiles. Nucleic Acids Res. **41**(D1), D320–D326 (2013)
21. Holm, L., Kaariainen, S., Rosenstrom, P., Schenkel, A.: Searching protein structure databases with DaliLite v. 3. Bioinformatics **24**, 2780–2781 (2008)
22. Hong, H.J., Chen, P.Y., Shih, T.C., Ou, C.Y., Jhuo, M.D., Huang, Y.Y., Cheng, C.H., Wu, Y.C., Chung, J.G.: Computational pharmaceutical analysis of anti-Alzheimer's Chinese medicine Coptidis Rhizoma alkaloids. Mol. Med. Rep. **5**(1), 142–147 (2012)
23. Jmol Homepage: Jmol: an Open-Source Java Viewer for Chemical Structures in 3D. http://www.jmol.org
24. Källberg, M., Wang, H., Wang, S., Peng, J., Wang, Z., Lu, H., Xu, J.: Template-based protein structure modeling using the RaptorX web server. Nat. Protoc. **7**, 1511–1522 (2012)
25. Kelley, L.A., Sternberg, M.J.E.: Protein structure prediction on the web: a case study using the Phyre server. Nat. Protoc. **4**(3):363–371 (2009)
26. Kim, D.E., Chivian, D., Baker, D.: Protein structure prediction and analysis using the Robetta server. Nucleic Acids Res. **32**(Suppl 2), W526–W531 (2004)
27. Laskowski, R.A., MacArthur, M.W., Moss, D.S., Thornton, J.M.: PROCHECK: a program to check the stereochemical quality of protein structures. J. Appl. Cryst. **26**, 283–291 (1993)
28. Leach, A.: Molecular Modelling: Principles and Applications, 2nd edn. Pearson Education EMA, London (2001)
29. Leaver-Fay, A., Tyka, M., Lewis, S.M., Lange, O.F., Thompson, J., Jacak, R., et al.: ROSETTA3: an object-oriented software suite for the simulation and design of macromolecules. Methods Enzymol. **487**, 545–574 (2011)
30. Małysiak, B., Momot, A., Kozielski, S., Mrozek, D.: On using energy signatures in protein structure similarity searching. In: Rutkowski, L., et al. (eds.) AISC 2008, Lecture Notes Computer Science, vol. 5097, pp. 939–950. Springer, Heidelberg (2008)
31. Mrozek, D., Małysiak, B., Kozielski, S.: An optimal alignment of proteins energy characteristics with crisp and fuzzy similarity awards. In: Proceedings of the IEEE International Conference on Fuzzy Systems (FUZZ-IEEE), pp. 1508–1513 (2007)
32. Mrozek, D., Małysiak, B., Kozielski, S.: EAST: energy alignment search tool. In: Wang, L., et al. (eds.): Proceedings of the 3rd IEEE International Conference on Fuzzy Systems and Knowledge Discovery. Xi'an, China, Lecture Notes Computer Science, vol. 4223, pp. 696–705. Springer, Berlin (2006)
33. Mrozek, D., Małysiak, B., Kozielski, S.: Energy profiles in detection of protein structure modifications. In: Proceedings of the IEEE International Conference on Computing and Informatics, Kuala Lumpur, pp. 1–6 (2006)
34. Mrozek, D., Małysiak, B., Kozielski, S.: Energy properties of protein structures in the analysis of the human RAB5A cellular activity. Adv. Intell. Soft Comput. **59**, 121–131 (2009)
35. Mrozek, D., Małysiak-Mrozek, B., Kozielski, S., Górczynska-Kosiorz, S.: The EDML format to exchange energy profiles of protein molecular structures. Lecture Notes Computer Science, vol. 5754, Springer, pp. 146–157 (2009)
36. Mrozek, D., Małysiak-Mrozek, B., Kozielski, S., Świerniak, A.: The Energy Distribution Data Bank: collecting energy features of protein molecular structures. In Proceedings of the 9th IEEE International Conference on Bioinformatics and Bioengineering, IEEE, pp. 1–6 (2009)
37. Mrozek, D., Małysiak-Mrozek, B., Kozielski, S.: Alignment of protein structure energy patterns represented as sequences of fuzzy numbers. In: Fuzzy Information Processing Society, 2009. NAFIPS 2009. Annual Meeting of the North American Fuzzy Information Processing Society, pp. 1–6 (2009)
38. Mrozek, D., Małysiak-Mrozek, B.: An improved method for protein similarity searching by alignment of fuzzy energy signatures. Int. J. Comput. Intell. Syst. **4**(1):75–88 (2011)
39. Mrozek, D., Małysiak-Mrozek, B.: CASSERT: a two-phase alignment algorithm for matching 3D structures of proteins. In: Kwiecień, A., Gaj, P., Stera, P. (eds.) CN 2013, CCIS, vol. 370, pp. 334–343 (2013)

40. Mrozek, D., Mastej, A., Małysiak, B.: Protein molecular viewer for visualizing structures stored in the PDBML format. In: Pietka, E., Kawa, J. (eds.) Information Technologies in Biomedicine, AISC, vol. 47, pp. 377–386. Springer, Berlin (2008)
41. Mrozek, D., Wieczorek, D., Małysiak-Mrozek, B., Kozielski, S.: PSS-SQL: protein secondary structure–structured query language. In: Proceedings of 32th Annual International Conference of the IEEE Engineering in Medicine and Biology Society, EMBS 2010, Buenos Aires, Argentina, pp. 1073–1076 (2010)
42. Mrozek, D., Małysiak-Mrozek, B., Kozielski, S.: Protein comparison by the alignment of fuzzy energy signatures. RSKT 2009. Lect. Notes Comput. Sci. **5589**, 289–296 (2009)
43. Mrozek, D., Brożek, M., Małysiak-Mrozek, B.: Parallel implementation of 3D protein structure similarity searches using a GPU and the CUDA. J. Mol. Model **20**, 2067 (2014)
44. Needleman, S.B., Wunsch, C.D.: A general method applicable to the search for similarities in the amino acid sequence of two proteins. J. Mol. Biol. **48**(3), 443–453 (1970)
45. Oostenbrink, C., Villa, A., Mark, A.E., van Gunsteren, W.: A biomolecular force field based on the free enthalpy of hydration and solvation: the GROMOS force-field parameter sets 53A5 and 53A6. J. Comp. Chem. **25**, 1656–1676 (2004)
46. Pearson, W.R.: Flexible sequence similarity searching with the FASTA3 program package. Methods Mol. Biol. **132**, 185–219 (2000)
47. Ramachandran, G.N., Ramakrishnan, C., Sasisekaran, V.: Stereochemistry of polypeptide chain configurations. J. Mol. Biol. **7**, 95–99 (1963)
48. Rost, B., Liu, J.: The predict protein server. Nucleic Acids Res. **31**(13), 3300–3304 (2003)
49. Sayle, R.: RasMol, molecular graphics visualization tool. Biomolecular Structures Group, Glaxo Welcome Research & Development, Stevenage, Hartfordshire, 5/02/2013 (1998). http://www.umass.edu/microbio/rasmol/
50. Schrödinger, L.L.C.: The PyMOL molecular graphics system, version 1.3r1 (2010). http://www.pymol.org
51. Schulz, G.E., Schirmer, R.H.: Principles of Protein Structure. Springer, New York (1979)
52. Shapiro, J., Brutlag, D.: FoldMiner and LOCK2: protein structure comparison and motif discovery on the web. Nucleic Acids Res. **32**, 536–541 (2004)
53. Shindyalov, I., Bourne, P.: Protein structure alignment by incremental combinatorial extension (CE) of the optimal path. Protein Eng. **11**(9), 739–747 (1998)
54. Smith, T., Waterman, M.: Identification of common molecular subsequences. J. Mol. Biol. **147**, 195–197 (1981)
55. Smith, G.D., Dodson, G.G.: The structure of a rhombohedral R6 insulin hexamer that binds phenol. Biopolymers **32**(4), 441–445 (1992)
56. Söding, J.: Protein homology detection by HMM-HMM comparison. Bioinformatics **21**, 951–960 (2005)
57. Stanek, D., Mrozek, D., Małysiak-Mrozek, B.: MViewer: visualization of protein molecular structures stored in the PDB, mmCIF and PDBML data formats. In: Kwiecień, A., Gaj, P., Stera, P. (eds.) CN 2013, CCIS vol. 370, pp. 323–333 (2013)
58. Taylor, W.R., Orengo, C.A.: A local alignment method for protein structure motifs. J. Mol. Biol. **233**, 488–497 (1993)
59. Warecki, S., Znamirowski, L.: Random simulation of the nanostructures conformations. In: Proceedings of International Conference on Computing, Communication and Control Technology, vol. 1, The International Institute of Informatics and Systemics, Austin, Texas, pp. 388–393 (2004)
60. Watson, H.: The stereochemistry of the protein myoglobin. Prog. Stereochem. **24**, 299 (1969)
61. Wu, S., Skolnick, J., Zhang, Y.: Ab initio modeling of small proteins by iterative TASSER simulations. BMC Biol. **5**, 17 (2007)
62. Xu, J., Li, M., Kim, D., Xu, Y.: RAPTOR: optimal protein threading by linear programming, the inaugural issue. J. Bioinform. Comput. Biol. **1**(1), 95–117 (2003)
63. Xu, D., Zhang, Y.: Ab initio protein structure assembly using continuous structure fragments and optimized knowledge-based force field. Proteins **80**(7), 1715–1735 (2012)

64. Yang, Y., Faraggi, E., Zhao, H., Zhou, Y.: Improving protein fold recognition and template-based modeling by employing probabilistic-based matching between predicted one-dimensional structural properties of the query and corresponding native properties of templates. Bioinformatics **27**, 2076–2082 (2011)
65. Ye, Y., Godzik, A.: Flexible structure alignment by chaining aligned fragment pairs allowing twists. Bioinformatics **19**(2), 246–255 (2003)
66. Zhu, J., Weng, Z.: FAST: a novel protein structure alignment algorithm. Proteins **58**, 618–627 (2005)
67. Znamirowski, L.: Non-gradient, sequential algorithm for simulation of nascent polypeptide folding. Computational Science ICCS 2005. Lecture Notes in Computer Science vol. 3514, pp. 766–774 (2005)
68. Znamirowski, L.: Switching. VLSI Structures, Reprogrammable FPAA Structures, Nanostructures. Studia Informatica, vol. 25, no. 4A (60), Gliwice, pp. 1–236 (2004)

Chapter 2
Multithreaded PSS-SQL for Searching Databases of Secondary Structures

...; life was no longer considered to be a result of mysterious and vague phenomena acting on organisms, but instead the consequence of numerous chemical processes made possible thanks to proteins.

Amit Kessel, Nir Ben-Tal, 2010 [13]

Abstract Protein secondary structure (PSS), as an organizational level, provides important information regarding protein construction and regular spatial shapes, including alpha-helices, beta-strands, and loops, which protein amino acid chain can adopt in some of its regions. The relevance of this information and the scope of its practical applications cause the requirement for its effective storage and processing. In this chapter, we will see how PSSs can be stored in the relational database and processed with the use of the protein secondary structure-structured query language (PSS-SQL). The PSS-SQL is an extension to the SQL language. It allows formulation of queries against a relational database in order to find proteins having secondary structures similar to the structural pattern specified by a user. In this chapter, we will see how this process can be accelerated by parallel implementation of the alignment using multiple threads working on multiple-core CPUs.

Keywords Proteins · Secondary structure · Query language · SQL · Relational database · Multithreading · Parallel computing · Alignment

2.1 Introduction

Secondary structures are a kind of intermediate organizational level of protein structures, a level between the simple amino acid sequence and complex 3D structure. The analysis of protein structures on the basis of the secondary structures is very supportive for many processes that are important from the viewpoint of biomedicine

D. Mrozek, *High-Performance Computational Solutions in Protein Bioinformatics*, SpringerBriefs in Computer Science, DOI: 10.1007/978-3-319-06971-5_2,

and pharmaceutical industry, e.g., drug design. Algorithms comparing protein 3D structures and looking for structural similarities quite often make use of the secondary structure representation at the beginning as one of the features distinguishing one protein from the other. Secondary structures are taken into account in algorithms, such as VAST [8], LOCK2 [20], CTSS [5], CASSERT [16]. Also in protein 3D structure prediction by comparative modeling [12, 28], particular regions of protein structures are modeled through the adoption of particular secondary structure types of proteins that structure is already determined and deposited in a database. Secondary structure organizational level also shows what types of secondary structure a protein molecule is composed of, what is their arrangement—whether they are segregated or alternating each other. Based on the information proteins are classified by systems, such as CATH [19] and SCOP [18]. All these examples show how important the description by means of secondary structures is.

For scientists studying structures and functions of proteins, it is very important to collect data describing protein construction in one place and have the ability to search particular structures that satisfy given searching criteria. Consequently, this needs an appropriate representation of protein structures allowing for effective storage and searching. The problem is particularly important in the face of dynamically growing amount of biological and biomedical data in databases, such as PDB [4] or Swiss-Prot [3].

At the current stage of development of IT technologies, a well-established position in terms of collecting and managing various types of data reached relational databases [6]. Relational databases collect data in tables (describing part of reality) where data are arranged in columns and rows. Modern relational databases also provide a declarative query language—SQL that allows retrieving and processing collected data. The SQL language gained a great power in processing regular data hiding details of the processing under a quite simple SELECT statement. However, processing biological data, such as protein secondary structures (PSSs), by means of relational databases are hindered by several factors:

- Data describing protein structures have to be managed by database management systems (DBMSs), which work excellent in commercial uses, but they are not dedicated for storing and processing biological data. They do not provide the native support for processing biological data with the use of the SQL language, which is a fundamental, declarative way of data manipulation in most modern relational database systems.
- Processing of biological data must be performed by external tools and software applications, forming an additional layer in the IT system architecture, which is a disadvantage.
- Currently, results of data processing are returned in different formats, like: table-form datasets, TXT, HTML, or XML files, and users must adopt them in their software applications.
- Secondary processing of the data is difficult and requires additional external tools.

In other words, modern relational databases require some enhancements in order to deal with the data on secondary structures of proteins. The possibility of collecting

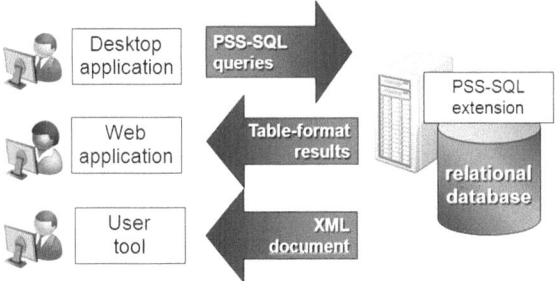

Fig. 2.1 Exploration of protein secondary structures in relational databases using PSS-SQL language. Secondary structure description of protein molecules is stored in relational database. The database management system (DBMS) has the PSS-SQL extension that interprets queries submitted by users. Users can connect to the database from various tools, desktop software applications, and Web applications. They obtain results of their queries in a table-like format or as an XML document

protein structural data in appropriate manner and processing the data by submitting simple queries to a database simplifies a work of many researchers working in the area of protein bioinformatics. Actually, the problem of storing biological data describing biopolymer structures of proteins and DNA/RNA molecules and possessing appropriate query language allowing processing the data has been noticed in the last decade and reported in several papers. There are only a few initiatives in the world reporting this kind of solutions.

For example, the ODM BLAST [23] is a successful implementation of the BLAST family of methods in the commercial Oracle database management system. ODM BLAST extends the SQL language by providing appropriate functions for local alignment and similarity searching of DNA/RNA and protein amino acid sequences. ODM BLAST works fast, but in terms of protein molecules it is limited only to the primary structure. In [9], authors describe their extension to the SQL language, which allows searching on the secondary structures of protein sequences. The extension was developed in Periscope (dedicated engine) and in Oracle (commercial database system). In the solution, secondary structures are represented by segments of different types of secondary structure elements (SSEs), e.g., hhhllleee. In [24], authors show the Periscope/SQ extension of the Periscope system. Periscope/SQ is a declarative tool for querying primary and secondary structures. To this purpose authors introduced new language PiQL, new data types, and algebraic operators according to the defined query algebra PiOA. The PiQL language has many possibilities. In this paper [25], the authors present their extensions to the object-oriented database (OODB) by adding the Protein-QL query language and the Protein-OODB middle layer for requests submitted to the OODB. Protein-QL allows to formulate simple queries that operate on the primary, secondary, and tertiary level.

Finally in 2010, me and a group of researchers from my university (Silesian University of Technology in Gliwice, Poland) developed the PSS-SQL [15, 17, 26, 27], which is an extension to the Transact-SQL language and Microsoft SQL Server DBMS allowing for searching protein similarities on the secondary structure level (Fig. 2.1).

I had the opportunity to be the manager and supervisor of the project, and I have never stopped thinking on its improvement in the following years. New versions of the PSS-SQL consists of many improvements leading to the significant growth of the efficiency of PSS-SQL queries, including:

- parallel and multithreaded execution of the alignment procedure used in the searching process,
- reduction of the computational complexity of the alignment algorithm by using gap penalty matrices, and
- indexing of sequences of SSEs.

The PSS-SQL language containing these improvements will be described in this chapter. In the chapter, we will also see results of performance tests for sample queries in PSS-SQL language and how to return query results as table-like result sets and as XML documents.

2.2 Storing and Processing Secondary Structures in a Relational Database

Searching for protein similarities on secondary structures by formulating queries in PSS-SQL requires that data describing secondary structures should be stored in a database in an appropriate format. The format should guarantee an efficient processing of the data. In PSS-SQL the search process is carried out in two phases, by:

1. Multiple scanning of a dedicated Segment Index for secondary structures.
2. Alignment of found segments in order to return k-best solutions.

All these steps, including data preparation, creating and scanning the Segment Index, and alignment will be discussed in the following sections.

2.2.1 Data Preparation and Storing

The PSS-SQL uses a specific representation of PSSs while storing them in a database.

Let us assume, we have a protein P described by the amino acid sequence (primary structure):

$$P = \{p_i | i = 1, 2, \ldots, n \land p_i \in \Pi \land n \in \mathbb{N}\}, \tag{2.1}$$

where n is the length of protein amino acid chain, i.e., the number of amino acids, and Π is a set of 20 common types of amino acids.

Secondary structure of protein P can be then described as a sequence of SSEs related to amino acids in the protein chain:

$$S = \{s_i | i = 1, 2, \ldots, n \land s_i \in \Sigma \land n \in \mathbb{N}\}, \tag{2.2}$$

```
P0A2U6
ADCC_STRPN
Zinc transport system ATP-binding protein adcC OS=Streptococcus pneumoniae GN=adcC PE=3 SV=1

MRYITVEDLSFYYDREPVLEHINYCVDSGEFVTLTGENGAAKTTLIKASLGILQPRIGRVAISKTNTQGKKLRIAYLPQQIASFNAGFPSTV
YEFVKSGRYPRKGWFRRLNAHDEEHIKASLDSVGMWEHRDKRLGSLSGGQKQRAVIARMFASDPDVFILDEPTTGMDAGSKNEFYELMHHSA
HHHGKAVLMITHDPEEVKDYADRNIHLVRNQDSPWRCFNVHENGQEVGHA


CCCEEECCCEEECCCCCCEEEEEEECCCCCEEEECCCCCCCHHHHHEEEECCCCCCCCEEEEECCCCCCEEEEEEHHHHHHHHHCCCCCCE
EEEECCCCCCCHHHHHCCCCCHHHHHHHHHHHHCCCCCCCCCCCCCCCCCHHHHHHHHHHHHCCCCEEEECCCCCCCCCCCCHHHHHHHHCC
CCCCEEEEEEECCCCCCCCCCCCCCEEEEECCCCCCEEEECCCCCCCCCC
```

Fig. 2.2 Sample amino acid sequence of *Zinc transport system ATP-binding protein adcC* in the *Streptococcus pneumoniae* with the corresponding sequence of secondary structure elements

```
id    protID       protAC  name                     length primary                           secondary
----- ------------ ------- ------------------------ ------ --------------------------------- -----------------------------
3294  2DRA_HUMAN   P01903  HLA class II histoco...  254    MAISGVPVLGFFIIAVLMSAQESWAIK...    CCCCCEEECCCEEEHHHHHHHHHHHH...
3296  2ENR_CLOTY   P11887  2-enoate reductase (...  30     MKNKSLFEVIKIGKVEVXXKIXMAVMG...    CCCCCCEEEEECCCCCCCCCCCCEEEE...
3297  2NPD_NEUCR   Q01284  2-nitropropane dioxy...  378    MHFPGHSSKKEESAQAALTKLNSWFPT...    CCCCCCCCCHHHHHHHHHHHCCCCCC...
3299  2PS_GERHY    P48391  2-pyrone synthase OS...  402    MGSYSSDDVEVIREAGRAQGLATILAI...    CCCCCCCEEEEEEHHHHHHHHHEEEEE...
3300  ACSA1_PSEPK  Q88EH6  Acetyl-coenzyme A sy...  653    MSAAPLYPVRPEVAATTLTDEATYKAM...    CCCEEEEECCHHHHHHHHHHHHHHHH...
3302  ACSA2_ACEXY  Q59167  Cellulose synthase 2...  1596   MIYRAILKRLRLEQLARVPAVSAASPF...    CCCHHHHHHHHHHHHHHHHCEEECCCCE...
```

Fig. 2.3 Sample relational table storing sequences of secondary structure elements (SSEs) (*secondary* field), amino acid sequences (*primary* field), and additional information of proteins from the Swiss-Prot database. The table (called *ProteinTbl*) will be used in sample queries presented in next sections. Secondary structures were predicted from amino acid sequences using the Predator program [7]

where each element s_i corresponds to a single element p_i, and Σ is a set of secondary structure types. The set Σ may be defined in various ways. A widely accepted definition of the set provides DSSP [10, 11]. The DSSP code distinguishes the following secondary structure types:

- H = alpha helix,
- B = residue in isolated beta-bridge,
- E = extended strand, participates in beta ladder,
- G = 3-helix (3/10 helix),
- I = 5 helix (pi helix),
- T = hydrogen bonded turn, and
- S = bend.

In practice, the set is often reduced to the three general types [7]:

- H = alpha helix,
- E = beta strand (or beta sheet), and
- C = loop, turn or coil.

An example of such a representation of protein structure is shown in Fig. 2.2, where we can see primary and secondary structures of a sample protein recorded as sequences. In such a way both sequences can be effectively stored in a relational database, as it is shown in Fig. 2.3.

Fig. 2.4 Part of the segment
table

id	protID	type	startPos	length
67	3	C	0	3
68	3	H	3	23
69	3	C	26	8
70	3	H	34	12
71	3	C	46	3
72	3	E	49	3

2.2.2 Indexing of Secondary Structures

At the level of DBMS, the PSS-SQL uses additional data structures and indexing in
order to accelerate the similarity searching. A dedicated *segment table* is created for
the table field storing sequences of secondary structures elements. The segment table
consists of secondary structures and their lengths extracted from the sequences of
SSEs, together with locations of the particular secondary structure in the molecule
(identified by the residue number, Fig. 2.4). Then, additional Segment Index is created
for the segment table. The Segment Index is a B-Tree clustered index holding on the
leaf level data pages from the additional segment table. The idea of using the segment
table and segment index is adopted from the work [9]. The Segment Index supports
preliminary filtering of protein structures that are not similar to the query pattern.
During the filtering, the PSS-SQL extension extracts the most characteristic features
of the query pattern and, on the basis of the information in the index, eliminates
proteins that do not meet the search criteria. Afterward, proteins that pass the filtering
process are aligned to the query pattern.

If we take a closer look at the segment table, we will see that it stores secondary
structures in the form that has been described in Sect. 1.3.2. During the scanning
of the Segment Index the search engine of the PSS-SQL tries to match segments
distinguished in the given query pattern to segments of the index.

2.2.3 Alignment Algorithm

The alignment implemented in the PSS-SQL is inspired by the Smith–Waterman
method [21]. The method allows to align two biopolymer sequences, originally
DNA/RNA sequences or amino acid sequences of proteins. When scanning a database the alignment is performed for each pair of sequences—query sequence given
by a user and a successive, qualified sequence from a database. In PSS-SQL, after performing multiple scanning of the Segment Index (MSSI), a database protein
structure S^D of the length d residues is represented as a sequence of segments (see
also formulas 1.14 and 1.15), which can be expanded to the following form:

$$S^D = SSE_1^D L_1, SSE_2^D L_2, \ldots, SSE_n^D L_n, \tag{2.3}$$

where $SSE_i^D \in \Sigma$ describes the type of secondary structure (as defined in Sect. 2.2.1), n is the number of segments (secondary structures) in a database protein, $L_i \le d$ is the length of the ith segment of a database protein S^D.

Query protein structure S^Q, given by a user in a form of string pattern, is represented by ranges, which gives more flexibility in defining search criteria against proteins in a database:

$$S^Q = SSE_1^Q(L_1; U_1), SSE_2^Q(L_2; U_2), \ldots, SSE_m^Q(L_m; U_m), \qquad (2.4)$$

where $SSE_j^Q \in \Sigma$ describes the type of secondary structure (as defined in Sect. 2.2.1), $L_j \le U_j \le q$ are lower and upper limits for the number of successive SSEs of the same type, q is the length of the query protein S^Q measured in residues, which is the maximal length of the string query pattern resulting from expanding the ranges of the pattern, m is the number of segments in the query pattern.

Additionally, the SSE_j^Q can be replaced by the wildcard symbol '?', which denotes any type of SSE from Σ, and the value of the U_j can be replaced by the wildcard symbol '*', which denotes $U_j = +\infty$.

The advantage of the used alignment method is that it finds local, optimal alignments with possible gaps between corresponding elements. A big drawback is that it is computationally costly, which negatively affects efficiency of the search process carried out against the whole database. The computational complexity of the original algorithm is $O(n * m(n + m))$ when allowing for gaps calculated in a traditional way. However, in the PSS-SQL we have modified the way how gap penalties are calculated, which results in better efficiency.

While aligning two protein structures S^Q and S^D, the search engine of the PSS-SQL calculates the similarity matrix D according to the following formulas.

$$D_{i,0} = 0 \quad \text{for} \quad i \in [0, q], \qquad (2.5)$$

and

$$D_{0,j} = 0 \quad \text{for} \quad j \in [0, d], \qquad (2.6)$$

and

$$D_{i,j} = max \begin{cases} 0 \\ D_{i-1,j-1} + d_{i,j} \\ E_{i,j} \\ F_{i,j} \end{cases}, \qquad (2.7)$$

for $i \in [1, q]$, $j \in [1, d]$, where q, d are lengths of proteins S^Q and S^D, and $d_{i,j}$ is the matching degree between elements $SSE_i^D L_i$ and $SSE_j^Q(L_j; U_j)$ of both structures calculated using the following formula:

$$d_{i,j} = \begin{cases} \omega_+ & \text{if } SSE_i^D = SSE_j^Q \wedge L_i \geq L_j \wedge L_i \leq U_j \\ \omega_- & \text{otherwise} \end{cases}, \qquad (2.8)$$

where ω_+ is the matching award, and ω_- is the mismatch penalty. If the element SSE_j^Q is equal to '?', then the matching procedure ignores the condition $SSE_i^D = SSE_j^Q$. Similarly, if we assign the '*' symbol for the U_j, the procedure ignores the condition $L_i \leq U_j$.

Auxiliary matrices E and F, called gap penalty matrices, allow to calculate horizontal and vertical gap penalties with the $O(1)$ computational complexity (as opposed to the original method, where it was possible with the $O(n)$ computational complexity for each direction). In the first version of the PSS-SQL, the calculation of the current element of the matrix D required an inspection of all previously calculated elements in the same row (for a horizontal gap) and all previously calculated elements in the same column (for a vertical gap). By using gap penalty matrices we need only to check one previous element in a row and one previous element in a column. Such an improvement gives a significant acceleration of the alignment method, and the acceleration is greater for longer sequences of SSEs and greater similarity matrices D. Elements of the gap penalty matrices E and F are calculated according to the following equations:

$$E_{i,j} = max \begin{cases} E_{i-1,j} - \delta \\ D_{i-1,j} - \sigma \end{cases}, \qquad (2.9)$$

and

$$F_{i,j} = max \begin{cases} F_{i,j-1} - \delta \\ D_{i,j-1} - \sigma \end{cases}, \qquad (2.10)$$

where σ is the penalty for opening a gap in the alignment, and δ is the penalty for extending the gap, and:

$$E_{i,0} = 0 \text{ for } i \in [0, q], \quad F_{i,0} = 0 \text{ for } i \in [0, q], \qquad (2.11)$$

$$E_{0,j} = 0 \text{ for } j \in [0, d], \quad F_{0,j} = 0 \text{ for } j \in [0, d]. \qquad (2.12)$$

The PSS-SQL uses the following values for matching award $\omega_+ = 4$, mismatch penalty $\omega_- = -1$, gap open penalty $\sigma = -1$, and gap extension penalty $\delta = -0.5$.

Filled similarity matrix D consists of many possible paths how two sequences of SSEs can be aligned. Backtracking from the highest scoring matrix cell and going along until a cell with score 0 is encountered allows to find the highest scoring alignment path. However, in the version of the alignment method that is implemented in the PSS-SQL, the search engine finds k-best alignments by searching consecutive maxima in the similarity matrix D. This is necessary, since the pattern is usually not defined precisely, contains ranges of SSEs or undefined elements. Therefore, there can be many regions in a protein structure that fit the pattern. In the process of finding alternative alignment paths, the alignment method follows the value of the internal parameter MPE (minimum path end), which defines the stop criterion. The search engine finds alignment paths until the next maximum in the similarity matrix D is lower than the value of the MPE parameter. The value of the MPE depends on the specified pattern, according to the following formula.

$$MPE = (MPL \times \omega_+) + (NoIS \times \omega_-), \qquad (2.13)$$

where MPL is the minimum pattern length, $NoIS$ is the number of imprecise segments, i.e., segments, for which $L_j \neq U_j$. For example, for the structural pattern $h(10;20)$, $e(1;10)$, $c(5)$, $e(5;20)$ containing α-helix of the length 10–20 elements, β-strand of the length 1–10 elements, loop of the length 5 elements, and β-strand of the length 5–20 elements, the $MPL = 21$ (10 elements of the type h, 1 element of the type e, 5 elements of the type c, and 5 elements of the type e), the $NoIS = 3$ (first, second, and fourth segment), and therefore, $MPE = 81$.

2.2.4 Multithreaded Implementation

In the original PSS-SQL [17], the calculation of the similarity matrix D was performed by a single thread. This negatively affected performance of PSS-SQL queries or, at least, this left a kind of computational reserve in the era of multicore CPUs. In the new version of the PSS-SQL we have reimplemented procedures and functions in order to use all processor cores that are available on the computer hosting the database with the PSS-SQL extension. A part of the work was carried out by B. Socha [22], my associate in this project.

However, the multithreaded implementation required different approach while calculating values of particular cells of the similarity matrix D. Successive cells cannot be calculated one by one, as in the original version, but calculations are carried out for cells located on successive diagonals, as it is shown in Fig. 2.5. This is because, according to Eqs. (2.7), (2.9), and (2.10) each cell $D_{i,j}$ can be calculated only if there are calculated cells $D_{i-1,j-1}$, $D_{i-1,j}$ and $D_{i,j-1}$. Such an approach to the calculation of the similarity matrix is called a *wavefront* [2, 14].

Moreover, in order to avoid too many synchronizations between running threads (which may lead to significant delays), the entire similarity matrix is divided to so-called areas (Fig. 2.6a). These areas are parts of the similarity matrix that have a smaller size $q' \times d'$. Assuming that the entire similarity matrix has the size of $q \times d$, where q and d are lengths of two compared sequences of SSEs, the number of areas that must be calculated is equal to:

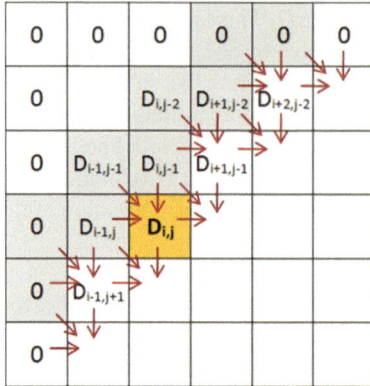

Fig. 2.5 Calculation of cells in the similarity matrix D by using the wavefront approach. Calculation is performed for cells at diagonals, since their values depend on previously calculated cells. *Arrows* show dependences of particular cells and the direction of value derivation

Fig. 2.6 Division of the similarity matrix D into areas (*left*)—arrows show mutual dependencies between areas during calculation of the matrix. (*right*) An order in which areas will be calculated in a sample similarity matrix

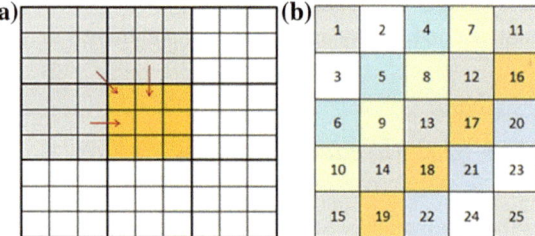

$$n_A = \left\lceil \frac{q}{q'} \right\rceil \times \left\lceil \frac{d}{d'} \right\rceil. \tag{2.14}$$

For example, for the matrix D of the size 382×108 and size of the area $q' = 10$ and $d' = 10$, the $n_A = \left\lceil \frac{382}{10} \right\rceil \times \left\lceil \frac{108}{10} \right\rceil = 39 \times 11 = 429$. Areas are assigned to threads working in the system. Each thread is assigned to one area, which is an atomic portion of calculation for the thread. Areas can be calculated according to the same wavefront paradigm. The area $A_{z,v}$ can be calculated, if there have been calculated areas $A_{z-1,v}$ and $A_{z,v-1}$ for $z > 0$ and $v > 0$, which implies an earlier calculation of the area $A_{z-1,v-1}$. The area $A_{0,0}$ is calculated as a first one, since there are no restrictions for calculation of the area.

In order to synchronize calculations, each area has a semaphore assigned to it. Semaphores guarantee that an area will not be calculated until the areas that it depends on have not been calculated. When all cells of an area have been calculated, the semaphore is being unlocked. Therefore, each area waits for unlocking two semaphores—for areas $A_{z-1,v}$ and $A_{z,v-1}$ for $z > 0$ and $v > 0$. While calculating an area each thread realizes the algorithm, which pseudocode is presented in Algorithm 1.

In Algorithm 1, after initialization of variables (lines 2–4), the thread enters the critical section marked with the *lock* keyword (line 5). Entering the critical section means that a thread obtains the mutual-exclusion lock for a given object. The thread executes some statements, and finally releases the lock. In our case, the thread obtains an exclusive access to the coordinates (z, v) of the area, which should be calculated by calling *GetAreaZ()* and *GetAreaV()* methods (lines 6–7). In the critical section, the thread also triggers the calculation of the (z, v) coordinates of the next area that should be calculated by another thread (line 8). Lines 9–11 determine whether this will be the last area that is calculated by any thread. Upon leaving the critical section, the current thread waits until areas $A_{z-1,v}$ and $A_{z,v-1}$ are unlocked (lines 13–14). Then, based on coordinates (z, v) and the area size in both dimensions, the thread determines absolute coordinates (i, j) of the first cell of the area (lines 15–16). These coordinates are used inside the following two *for* loops in order to establish absolute coordinates (i, j) of the current cell of the area. Figure 2.7 helps to interpret the variables used in the algorithm. The value of the current cell is calculated in line 21, according to formulas (2.5)–(2.7). When the thread completes the calculation of the current area, it unlocks the area (line 24) and asks for another area (lines 25–27).

Algorithm 1 The algorithm for the calculation of an area by a thread

---

1: **procedure** CALCULATEAREA
2:     $z \leftarrow 0$
3:     $v \leftarrow 0$
4:     $bool\,Finish \leftarrow true$
5:     **lock**   ▷ starts critical section
6:         $z \leftarrow GetAreaZ()$
7:         $v \leftarrow GetAreaV()$
8:         Calculate (z, v) coordinates of the next area
9:         **if** calculation successful (i.e., exists next area) **then**
10:            $bool\,Finish \leftarrow false$
11:        **end if**
12:    **endlock**
13:    Wait for unlocking the area $A_{z-1,v}$
14:    Wait for unlocking the area $A_{z,v-1}$
15:    $absStart\_i \leftarrow z * areaSizeZ$
16:    $absStart\_j \leftarrow v * areaSizeV$
17:    **for** $rel\_i \leftarrow 0$ **to** $areaSizeZ - 1$ **do**
18:        **for** $rel\_j \leftarrow 0$ **to** $areaSizeV - 1$ **do**
19:            $i \leftarrow absStart\_i + rel\_i$
20:            $j \leftarrow absStart\_j + rel\_j$
21:            Calculate cell $D_{i,j}$ according to formulas 2.5-2.7
22:        **end for**
23:    **end for**
24:    Unlock area $A_{z,v}$
25:    **if** $\neg bool\,Finish$ **then**
26:        Apply for the next area (enqueue for execution)
27:    **end if**
28: **end procedure**

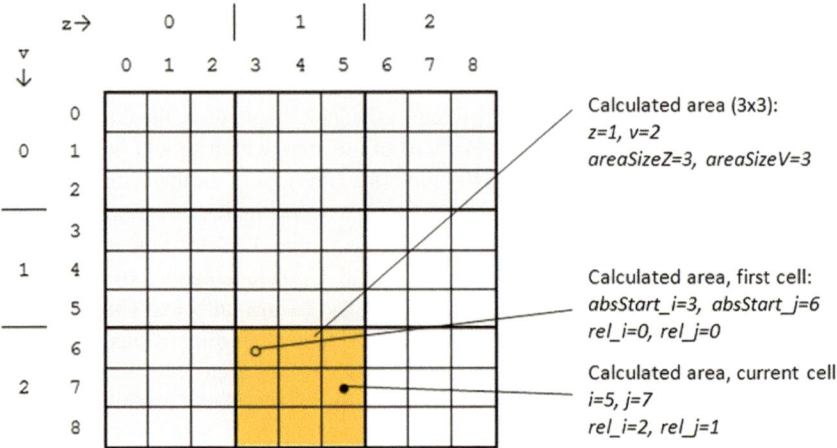

Fig. 2.7 Interpretation of variables used in the Algorithm 1 for the calculated area

The order in which areas are calculated is provided by a scheduling algorithm dispatching areas to threads. For example, the order of calculation particular areas in similarity matrix of the size 5×5 areas is shown in Fig. 2.6b. Such a division of the similarity matrix into areas reduces the number of tasks related to initialization of semaphores needed for synchronization purposes and reduces the synchronization time itself, which increases the efficiency of the alignment algorithm. For the PSS-SQL, the size of the area was set to 3×7 elements (3 for query protein, 7 for database protein) on the basis of experiments conducted by Socha [22].

2.3 SQL as the Interface Between User and the Database

PSS-SQL extends the standard syntax of the SQL language by providing additional functions that allow to search protein similarities on secondary structures. SQL language becomes a user interface (UI) between the user, who is a data consumer, and DBMS hosting secondary structures of proteins. PSS-SQL discloses three important functions for scanning PSSs: *containSequence*, *sequencePosition*, and *sequenceMatch*; all will be described in this chapter. PSS-SQL covers also a series of supplementary procedures and functions, which are used implicitly, e.g., for extracting segments of particular types of SSEs, building additional segment tables, indexing SSEs sequences, processing these sequences, aligning the target structures from a database to the query pattern, validating patterns, and many other operations. PSS-SQL extension was developed in the C# programming language. All procedures were assembled in the ProteinLibrary DLL file and registered for the Microsoft SQL Server 2008R2/2012 (Fig. 2.8).

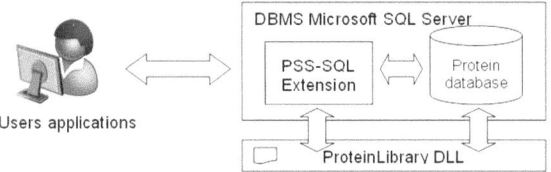

Fig. 2.8 General architecture of the system with the PSS-SQL extension. The PSS-SQL extension is registered in the Microsoft SQL Server DBMS. When the user submits a query invoking PSS-SQL functions (actually, Transact-SQL functions) the DBMS redirects the call to the PSS-SQL extension, which invokes appropriate functions assembled in the ProteinLibrary DLL library, passing appropriate parameters

2.3.1 Pattern Representation in PSS-SQL Queries

While searching protein similarities on secondary structures, we need to pass the query structure (query pattern) as a parameter of the search procedure. In PSS-SQL queries the pattern is represented as in the formula (2.4). Such a representation allows users to formulate a large number of various query types with different degrees of complexity. Moreover, we assumed that query patterns should be as simple as possible and should not cause any syntax difficulties. Therefore, we have defined the corresponding grammar in order to help constructing the query pattern.

In simple words, in PSS-SQL queries, the pattern is represented by blocks of segments. Each segment is determined by its type and length. The segment length can be represented precisely or as an interval. It is possible to define segments, for which the type is not important or undefined (wildcard symbol '?'), and for which the upper limit of the interval is not defined (wildcard symbol '*'). The grammar for defining patterns written in the Chomsky notation has the following form. The grammar is formally defined as the ordered quad-tuple:

$$G_{pss} = \langle N_{pss}, \Sigma_{pss}, P_{pss}, S_{pss} \rangle, \tag{2.15}$$

```
Σpss = {c, h, e, ?, *, N₊}
Npss = { <sequence>, <blocks_of_segments>, <segment>, <type>, <begin>,
 <end>, <length>, <integer_greater_than_zero_and_zero>, <undetermined> }
Ppss = {
 <sequence> ::= <blocks_of_segments>
 <blocks_of_segments> ::= <segment> | <segment>, <blocks_of_segments>
 <segment> ::= <type> (<begin>; <end>) | <type> (<length>)
 <begin> ::= <integer_greater_than_zero_or_zero>
 <end> ::= <integer_greater_than_zero_or_zero> | <undetermined>
 <length> ::= <integer_greater_than_zero_or_zero>
 <type> ::= c | h | e | ?
 <integer_greater_than_zero_or_zero> ::= N₊ | 0
 <undetermined> ::= * }
Spss = <sequence>
```

where the symbols respectively mean: N_{pss}—a finite set of nonterminal symbols, Σ_{pss}—a finite set of terminal symbols, P_{pss}—a finite set of production rules, S_{pss}—a distinguished symbol $S \in N_{pss}$ that is the start symbol.

Assumption: `<begin>` $\leq$ `<end>`

The following terms are compliant with the defined grammar G_{pss}:

- `h(1;10)`—representing an α-helix of the length 1–10 elements;
- `e(2;5),h(10;*),c(1;20)`—representing a β-strand of the length 2–5 elements, followed by an α-helix of the length at least 10 elements, and a loop of the length 1–20 elements;
- `e(10;15),?(5;20),h(35)`—representing a β-strand of the length 10–15 elements, followed by any element of the length 5–20, and an α-helix of the exact length 35 elements.

With such a representation of the query pattern, we can start the search process using one of the functions disclosed by PSS-SQL extension.

2.3.2 Sample Queries in PSS-SQL

The PSS-SQL extension provides a set of functions and procedures for processing PSSs. Three of the functions can be effectively invoked from the SQL commands, usually the SELECT statement.

The *containSequence* function verifies if a particular protein or a set of database proteins contain the structural pattern specified as a query pattern. This function returns the Boolean value 1 (true), if the database protein contains specified pattern, or 0 (false), if the protein does not include the pattern.

Sample invocation of the function is shown in Listing 2.1.

```
1  SELECT protID, protAC
2  FROM ProteinTbl
3  WHERE name LIKE '%Escherichia coli%' AND
4      dbo.containSequence(id, 'secondary', 'h(5;15),c(3),?(6),c(1;5)')=1
```

Listing 2.1 Sample query invoking *containSequence* function and returning identifiers of proteins from *Escherichia coli* containing the given secondary structure pattern.

The sample query returns identifiers and accession numbers of proteins from *Escherichia coli* having the structural region containing an α-helix of the length 5–15 elements, 3-element loop, any structure of the length 6 elements, and a loop of the length up to 5 elements (pattern `h(5;15),c(3),?(6),c(1;5)`).

Partial results of the query from Listing 2.1 are shown below.

```
protID          protAC
------------    --------
ACTP_ECOUT      Q1R3J9
ADD_ECOLC       B1IQD2
ADD_ECOLI       P22333
ADEC_ECO24      A7ZTM0
ADEC_ECO57      Q7A9L5
...
```

The *containSequence* function can be used in the SELECT and the WHERE phrase of the SQL SELECT statement. It is also possible to use the function in the WHERE clause of other DML statements, including UPDATE and DELETE, if needed. Detailed description of input arguments of the *containSequence* function is given in Table 2.1.

The *sequencePosition* and *sequenceMatch* functions allow to match the specified pattern to the structure of a protein or a group of database proteins. Pattern searching and matching is performed by multiple scanning of the segment index built on the segment table, followed by the alignment of the found segments. Both functions return a table containing information about the location of query pattern in the structure of each database protein. Both functions differ in the way how they are invoked in PSS-SQL queries.

Sample queries invoking both functions are shown in Listing 2.2. The function accepts the same arguments according to the list presented in Table 2.1. Since they return a table of values, they are nested in the FROM clause of SQL statements (mainly SELECTs, but also possible in some variants of UPDATE and DELETE statements). The use of the CROSS APPLY operator, instead of traditional JOIN, allows to avoid specifying the join condition, shortens the query syntax and, what even more important, improve performance, in the case of complex filtering conditions in the WHERE clause.

```
1  – invoking sequenceMatch and CROSS APPLY
2  SELECT p.protAC AS AC,p.name, s.startPos, s.endPos, p.[primary],
3      s.matchingSeq, p.secondary
4  FROM ProteinTbl AS p CROSS APPLY dbo.sequenceMatch(p.id, 'secondary',
5      'e(1;10),c(0;5),h(5;6),c(0;5),e(1;10),c(5)') AS s
6  WHERE p.name LIKE '%Staphylococcus aureus%' AND p.length > 150
7  ORDER BY AC, s.startPos
8
9  – invoking sequencePosition and standard JOIN
10 SELECT p.protAC AS AC, p.name, s.startPos, s.endPos, p.[primary],
11     s.matchingSeq, p.secondary
12 FROM ProteinTbl AS p JOIN dbo.sequencePosition('secondary',
13     'e(1;10),c(0;5),h(5;6),c(0;5),e(1;10),c(5)',
14     'p.name LIKE ''%Staphylococcus aureus%'' AND p.length > 150') AS s
15     ON p.id=s.proteinId
16 ORDER BY AC, s.startPos
```

Listing 2.2 Sample query invoking *sequenceMatch* and *sequencePosition* table functions and returning information on proteins from *Staphylococcus aureus* having the length greater than 150 residues and containing the given secondary structure pattern.

These sample queries return Accession Numbers (AC) and names of proteins from *Staphylococcus aureus* having the length greater than 150 residues and structural region containing β-strand of the length from 1 to 10 elements, optional loop up to 5 elements, an α-helix of the length 5–6 elements, optional loop up to 5 elements, a β-strand of the length 1–10 elements and a 5 element loop—pattern e(1;10),c(0;5),h(5;6),c(0;5),e(1;10),c(5).

Partial results of the query from Listing 2.2 are shown in Fig. 2.9. Detailed description of the output fields of the *sequenceMatch* and *sequencePosition* functions is given in Table 2.2.

Table 2.1 Input arguments of PSS-SQL functions

Argument	Description
@proteinId[a]	Unique identifier of a protein in the database table that contains sequences of SSEs (e.g. *id* field in case of the *ProteinTbl*)
@columnSSeq	Database field containing sequences of SSEs of proteins (e.g. *secondary*)
@pattern	Query pattern represented by a set of segments, e.g., h(2;10), c(1;5),?(2;*)
@predicate[b]	An optional, simple, or complex filtering criteria that allow to limit the list of proteins that will be processed during the search, e.g.,: *length* < 150

[a]except *sequencePosition*
[b]only *sequencePosition*

```
AC      name                              startPos endPos primary                  matchingSeq              secondary
------  --------------------------------  -------- ------ ------------------------  ----------------------   ------------------------
Q2FJ31  Alcohol dehydrogenase OS=S...     177      199    MRAAVVTKDHKVSIEDKK...     eeeehhhhheeeeeeeccc...    CCCEEEECCCCCCCCCCCCC...
Q2FJ31  Alcohol dehydrogenase OS=S...     187      218    MRAAVVTKDHKVSIEDKK...     eeeeeeeccccchhhhhcc...    CCCEEEECCCCCCCCCCCCC...
Q2G0G1  Alcohol dehydrogenase OS=S...     187      218    MRAAVVTKDHKVSIEDKK...     eeeeeeeccccchhhhhcc...    CCCEEEECCCCCCCCCCCCC...
Q2YSX0  Alcohol dehydrogenase OS=S...     187      218    MRAAVVTKDHKVSIEDKK...     eeeeeeeccccchhhhhcc...    CCCEEEECCCCCEEECCCCC...
Q5HI63  Alcohol dehydrogenase OS=S...     187      218    MRAAVVTKDHKVSIEDKK...     eeeeeeeccccchhhhhcc...    CCCEEEECCCCEEEHHHHHH...
Q6G0J3  Alcohol dehydrogenase OS=S...     187      218    MRAAVVTKDHKVSIEDKK...     eeeeeeeccccchhhhhcc...    CCCEEEECCCCEEEHHHHHH...
Q7A742  Alcohol dehydrogenase OS=S...     187      218    MRAAVVTKDHKVSIEDKK...     eeeeeeeccccchhhhhcc...    CCCEEEECCCCEEEHHHHHH...
Q99W07  Alcohol dehydrogenase OS=S...     187      218    MRAAVVTKDHKVSIEDKK...     eeeeeeeccccchhhhhcc...    CCCEEEECCCCEEEHHHHHH...
```

Fig. 2.9 Partial results of the sample queries from Listing 2.2 returned as a relational table, returned fields: *AC*—accession number, *name*—molecule name, *startPos*, *endPos*—position, where the pattern starts and ends in the target protein from a database, *primary*—amino acid sequence of the database protein, *matchingSeq*—exact sequence of SSEs, which matches to the pattern defined in the query, *secondary*—sequence of secondary structure elements SSEs of the database protein

Table 2.2 Output table of *sequenceMatch* and *sequencePosition* functions

Field	Description
proteinId	Unique identifier of the protein that contains the specified pattern
startPos	Position, where the pattern starts in the target protein from a database
endPos	Position, where the pattern ends in the target protein from a database
length	Length of the segment that matches to the given pattern
matchingSeq	Exact sequence of SSEs, which matches to the pattern defined in the query

Results of the PSS-SQL queries are originally returned in a tabular form. However, by adding an extra FOR XML clause at the end of the SELECT statement, like in the example in Listing 2.3, produces results in the XML format that can be easily transformed to the HTML web page by using appropriate XSLT transformation file, and finally, published in the Internet. Partial results of the query from Listing 2.3 are shown in Fig. 2.10. An additional function—*superimpose*—that was used in the presented query (Listing 2.3) visualizes the alignment of the matched sequence and the database sequence of SSEs.

Fig. 2.10 Partial results of
the query from Listing 2.3

```
<proteins>
  <protein>
    <AC>Q2FJ31</AC>
    <name>Alcohol dehydrogenase OS=Staphylococcus
          aureus (strain USA300)...</name>
    <startPos>177</startPos>
    <endPos>199</endPos>
    <matchingSeq>eeeehhhhhheeeeeeeccccc</matchingSeq>
    <primary>MRAAVVTKDHKVSIEDKKLRALKPGEALVQTEYCGVCH
      TDLHVKNADFGDVTGVTLGHEGIGKVIEVAED... </primary>
    <alignment>CCCEEEECCCCCCCCCCCCCCCCCCCCCCEEEEECC...
      CCCCEEECCCCCCCCCeeeehhhhhheeeeeeeccccccHHHHHH...
    </alignment>
  </protein>
  ...
</proteins>
```

```
1  SELECT p.protAC AS AC, p.name, s.startPos, s.endPos, s.matchingSeq, p .[primary],  dbo.superimpose
          (s.matchingSeq, p.secondary) AS alignment
2  FROM ProteinTbl AS p CROSS APPLY dbo.sequenceMatch(p.id, 'secondary',
3     'e(1;10),c(0;5),h(5;6),c(0;5),e(1;10).c(5)') AS s
4  WHERE p.name LIKE '%Staphylococcus aureus%'
5     AND p.length > 150
6  ORDER BY AC, s.startPos
7  FOR XML RAW ('protein'), ROOT('proteins'), ELEMENTS
```

Listing 2.3 Sample query invoking *sequenceMatch* table function and returning results as an XML
document by using the FOR XML clause.

2.4 Efficiency of the PSS-SQL

The efficiency of the PSS-SQL query language was examined in various experiments.
Tests were performed on the Microsoft SQL Server 2012 Enterprise Edition working
on nodes of the virtualized cluster controlled by the HyperV hypervisor hosted on
Microsoft Windows 2008 R2 Datacenter Edition 64-bit. The host server had the
following parameters: 2x Intel Xeon CPU E5620 2.40 GHz, RAM 32 GB, 3x HDD
1TB 7200 RPM. Cluster nodes were configured to use 4 CPU cores and 4GB RAM
per node, and worked under the Microsoft Windows 2008 R2 Enterprise Edition
64-bit operating system.

Most of the tests were performed on the database storing 6,360 protein structures.
However, in order to compare our language to one of the competitive solutions, some
tests were performed on the database storing 248,375 protein structures.

During the experiments, we measured execution times for various query patterns.
The query patterns were passed as a parameter of the *sequencePosition* function.
Tests were performed for queries containing the following sample patterns:

- SSE1: e(4;20),c(3;10),e(4;20),c(3;10),e(15),c(3;10),e(1;10)
- SSE2: h(30;40),c(1;5),?(50;60),c(5;10),h(29),c(1;5),h(20;25)
- SSE3: h(10;20),c(1;10),h(243),c(1;10),h(5;10),c(1;10),h(10;15)
- SSE4: e(1;10),c(1;5),e(27),h(1;10),e(1;10),c(1;10),e(5;20)
- SSE5: e(5;20),h(2;5),c(2;40),?(1;30),e(5;*)

Fig. 2.11 Execution time for various query patterns SSE1–SSE4 and for three variants of the PSS-SQL language: without multithreading (−MT), with multithreading, but without multiple scanning of the Segment Index (+MT−MSSI), with multithreading and with multiple scanning of the Segment Index (+MT+MSSI)

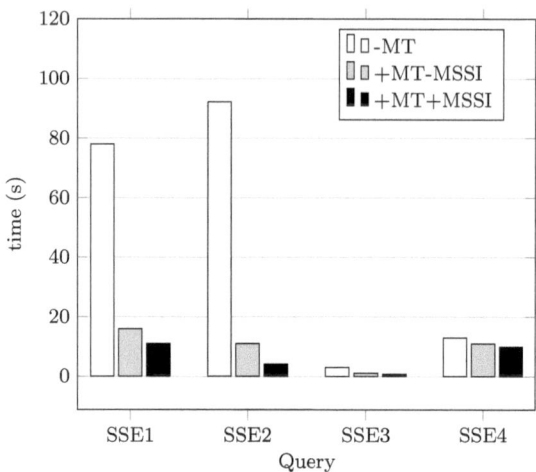

Pattern SSE1 represents protein structure built only with β-strands connected by loops. Pattern SSE2 consists of several α-helices connected by loops and one undefined segment of SSEs ('?' wildcard symbol). Patterns SSE3 and SSE4 have regions that are unique in the database, i.e., h(243) in pattern SSE3 and e(27) in pattern SSE4. Pattern SSE5 has a wildcard symbol '*' for undetermined length, which slows down the search process.

In order to verify the influence of particular acceleration techniques on the execution times, tests were carried out for the PSS-SQL in three variants:

- without multithreading (−MT),
- with multithreading, but without MSSI (+MT−MSSI), and
- with multithreading and with MSSI (+MT+MSSI).

Results of the tests shown in Fig. 2.11 prove that the performance of +MT−MSSI variant is higher, and in case of SSE1 and SSE2 even much higher, than −MT variant (implemented in original PSS-SQL). For +MT+MSSI we can see additional improvement of the performance. It is difficult to estimate the overall acceleration, because it tightly depends on the uniqueness of the pattern. The more unique the pattern is, the more proteins are filtered out based on the Segment Index, the fewer proteins are aligned and the less time we need to obtain results. We can see it clearly in Fig. 2.11 for patterns SSE3 and SSE4 that have precisely defined, unique regions h(243) and e(27). For universal patterns, like SSE1 and SSE2, for which we can find many fitting proteins or multiple alignments, we can observe longer execution times. In such cases, the parallelization and MSSI start playing a more significant role. In these cases, the length of the pattern influences the alignment time—for longer patterns we experienced longer response times. We have not observed any dependency between the type of the SSE and the response time.

However, specifying wildcards in the query pattern increases the waiting period, which is visible for the pattern SSE5 (Fig. 2.12). In Fig. 2.12 for the pattern SSE5, we

Fig. 2.12 Execution time for
query pattern SSE5 for three
variants of the PSS-SQL
language: without
multithreading (−MT),
with multithreading,
but without multiple
scanning of the Segment
Index (+MT−MSSI), with
multithreading and with
multiple scanning of the
Segment Index (+MT+MSSI)

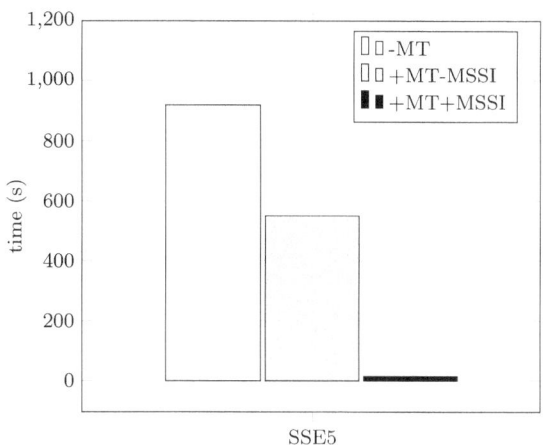

can also see how beneficial the use of the MSSI technique can be. In this particular
case, the execution time was reduced from 920 s in −MT (original PSS-SQL), and
550 s in +MT−MSSI, to 15 s in +MT+MSSI, which gives 61.33-times acceleration
over the −MT variant and 36.67-times acceleration over the +MT−MSSI variant.

2.5 Discussion

PSS-SQL language complements existing relational DBMSs, which are not designed
to process biological data, such as PSSs stored as sequences of SSEs. By extending
the standard SELECT, UPDATE, and DELETE statements of the SQL language,
it provides a declarative method for retrieving, modifying, and deleting records.
Records that satisfy the criteria given by a user can be returned in a table-like form
or as an XML document, which is easy to display as a Web page. In such a way, the
PSS-SQL extension to relational database management systems (RDBMS) provides
a kind of domain-specific language for processing PSSs. This is especially important for relational database designers, wide group of biological data analysts, and
bioinformaticians.

The PSS-SQL language can be used for the fast classification of proteins based
on their secondary structures. For example, systems such as SCOP [18] and CATH
[19] make use of the secondary structure description of protein structures in order to
classify proteins into classes and families. PSS-SQL can be also supportive in protein
3D structure prediction by homology modeling, where appropriate structure profile
can be found based on primary and secondary structure and the secondary structure
can be superimposed on the protein of the unknown 3D structure before performing
a free energy minimization.

Comparing the PSS-SQL to other languages presented in Sect. 2.1, we can notice that all variants of the PSS-SQL extend the syntax of the SQL. This makes the PSS-SQL similar to PiQL [24], rather than to ProteinQL [25]. ProteinQL was developed for the OODB and relies on its own domain-specific database and dedicated ProteinQL interpreter and translator. As opposed to ProteinQL, both PiQL, and PSS-SQL extend capabilities of RDBMS. They extend the syntax of the SQL language by providing additional functions that can be nested in particular clauses of the SQL commands. However, the form of queries provided by users is different. PiQL accepts query patterns in a full form, like in BLAST [1]—a tool used for fast local matching of biomolecular sequences of DNA and proteins. Query patterns provided in PSS-SQL are similar to those presented by Hammel and Patel in [9]. The pattern defined in a query does not have to be specified strictly. Segments in the pattern can be specified as intervals and they can have undefined lengths. Both languages allow specifying query patterns with undefined types of the SSE or patterns, where some SSE segments may occur optionally. Therefore, the search process has an approximate character, regarding various possible options for segment matching. The possibility of defining patterns that include optional segments allows users to specify gaps in a particular place.

The described version of the PSS-SQL also uses the method of scanning the Segment Index in order to accelerate the search process. The method was adopted from the work of Hammel and Patel [9]. However, after multiple scans of the Segment Index Hammel and Patel used sort-merge join operations in order to join segments from the same candidate proteins and decide, whether they meet specified query conditions or not. The novelty of PSS-SQL is that it relies on the alignment of the found segments. Alignment implemented in PSS-SQL gives the unique possibility of finding many matches for the same database protein and returning k-best matches, matches that in some particular cases can be separated by gaps. These are not the gaps defined by a user and specified by an optional segment, but the gaps providing better alignment of particular regions. This type of matching is typical for similarity searching between biomolecular sequences, such as DNA/RNA sequences or amino acid sequences. Presented approach extends the spectrum of searching and guarantees the optimality of the results according to assumed scoring system.

Despite the fact that PSS-SQL uses the alignment procedure, which is computationally complex, it gained quite a good performance. We have compared the efficiency of the PSS-SQL (+MT+MSSI variant) and language presented by Hammel and Patel for single-predicate exact match queries with various selectivity (between 0.3 and 6%) using the database storing 248,375 proteins (515 MB for *ProteinTbl*, 254 MB for segment table storing 11,986,962 segments). The PSS-SQL was on average 5.14 faster than *Comm-Seg* implementation, 3.28 faster than *Comm-CSP* implementation, both implemented on a commercial ORDBMS, and 1.84 faster than *ISS-MISS*(1) implementation on Periscope/SQ. This proves, that PSS-SQL compensates the efficiency loss caused by alignment procedure by using the Segment Index. In such a way, the PSS-SQL joins wide capabilities of the alignment process (possible gaps, mismatches, and many solutions), provides optimality and quality of results, and guarantees efficiency of scanning databases of secondary structures.

2.6 Summary

Integrating methods of PSS similarity searching with DBMSs provides an easy way for manipulation of biological data without the necessity of using external data mining applications. The PSS-SQL extension presented in this chapter is a successful example of such integration. PSS-SQL is certainly a good option for biological and biomedical data analysts who want to process their data on the server side. This has many advantages that are typical for such a processing in the client-server architecture. Entire logic of data processing is performed on the database server, which reduces the load on the user's computer. Therefore, data exploration is performed while retrieving data from a database. Moreover, the number of data returned to the user, and the network traffic between the server and the user application, are much reduced.

The use of multithreading allows to utilize the whole capable computing power more efficiently. The PSS-SQL adapts to the number of processing units possessed by the server hosting the DBMS and to the number of cores used by the database system. This results in better performance of the language while scanning huge databases of PSSs. For the latest information on the PSS-SQL, please visit the project home page: http://zti.polsl.pl/dmrozek/science/pss-sql.htm.

Parallelization of calculations in bioinformatics brings tangible benefits and reduces the execution time of many algorithms. In this chapter, we could see one of many examples of such parallelization. For readers that are interested in other examples I recommend the book *Parallel Computing for Bioinformatics and Computational Biology* by Zomaya [29] for further reading. In the next chapter, we will see how a massive parallelization of the 3D structure similarity searching on many-core CUDA-enabled GPU devices leads to reduction of the execution time of the process.

References

1. Altschul, S.F., Gish, W., Miller, W., Myers, E.W., Lipman, D.J.: Basic local alignment search tool. J. Mol. Biol. **215**, 403–410 (1990)
2. Anvik, J., MacDonald, S., Szafron, D., Schaeffer, J., Bromling, S., Tan, K.: Generating parallel programs from the wavefront design pattern. In: Proceedings of the 7th International Workshop on High-Level Parallel Programming Models and Supportive Environments (HIPS'02), Fort Lauderdale, Florida, April 2002, pp. 1–8 (2002)
3. Apweiler, R., Bairoch, A., Wu, C.H., et al.: Uniprot: the universal protein knowledgebase. Nucl. Acids Res. **32**(Database issue), D115–D119 (2004)
4. Berman, H., et al.: The Protein Data Bank. Nucl. Acids Res. **28**, 235–242 (2000)
5. Can, T., Wang, Y.: CTSS: a robust and efficient method for protein structure alignment based on local geometrical and biological features. In: Proceedings of the 2003 IEEE Bioinformatics Conference (CSB 2003), pp. 169–179 (2003)
6. Date, C.: An Introduction to Database Systems, 8th edn. Addison-Wesley, Reading (2003)
7. Frishman, D., Argos, P.: Incorporation of non-local interactions in protein secondary structure prediction from the amino acid sequence. Protein Eng. **9**(2), 133–142 (1996)
8. Gibrat, J., Madej, T., Bryant, S.: Surprising similarities in structure comparison. Curr. Opin. Struct. Biol. **6**(3), 377–385 (1996)

9. Hammel, L., Patel, J.M.: Searching on the secondary structure of protein sequences. In: Proceedings of 28th International Conference on Very Large Data Bases, Hong Kong, China, 2002, pp. 634–645 (2002)

10. Joosten, R.P., Te Beek, T.A.H., Krieger, E., Hekkelman, M.L., et al.: A series of PDB related databases for everyday needs. Nucl. Acid Res. **39**(Database issue), D411–D419 (2011)

11. Kabsch, W., Sander, C.: Dictionary of protein secondary structure: pattern recognition of hydrogen-bonded and geometrical features. Biopolymers **22**, 2577–2637 (1983)

12. Källberg, M., Wang, H., Wang, S., Peng, J., Wang, Z., Lu, H., Xu, J.: Template-based protein structure modeling using the RaptorX web server. Nat. Protoc. **7**, 1511–1522 (2012)

13. Kessel, A., Ben-Tal, N.: Introduction to Proteins: Structure, Function, and Motion, 1st edn. CRC Press, Boca Raton (2010)

14. Liu, W., Schmidt, B.: Parallel design pattern for computational biology and scientific computing applications. In: Proceedings of the 2003 IEEE International Conference on Cluster Computing, pp. 456–459 (2003)

15. Małysiak-Mrozek, B., Kozielski, S., Mrozek, D.: Server-side query language for protein structure similarity searching. In: Human-Computer Systems Interaction: Backgrounds and Applications. Springer, Berlin, Advances in Intelligent and Soft Computing **99**(2), 395–415 (2012)

16. Mrozek, D., Małysiak-Mrozek, B.: CASSERT: a two-phase alignment algorithm for matching 3D structures of proteins. In: Kwiecień, A., Gaj, P., Stera, P. (eds.) Proceedings of 22nd International Conference on Computer Networks, Communications in Computer and Information, Springer-Verlag, CCIS **370**, 334–343 (2013)

17. Mrozek, D., Wieczorek, D., Małysiak-Mrozek, B., Kozielski, S.: PSS-SQL: protein secondary structure—structured query language. In: Proceedings of 32nd Annual International Conference of the IEEE Engineering in Medicine and Biology Society, EMBS 2010, Buenos Aires, Argentina, pp. 1073–1076 (2010)

18. Murzin, A.G., Brenner, S.E., Hubbard, T., Chothia, C.: SCOP: a structural classification of proteins database for the investigation of sequences and structures. J. Mol. Biol. **247**, 536–540 (1995)

19. Orengo, C.A., Michie, A.D., Jones, S., Jones, D.T., et al.: CATH—a hierarchic classification of protein domain structures. Structure **5**(8), 1093–1108 (1997)

20. Shapiro, J., Brutlag, D.: FoldMiner and LOCK2: protein structure comparison and motif discovery on the web. Nucl. Acids Res. **32**, 536–541 (2004)

21. Smith, T., Waterman, M.: Identification of common molecular subsequences. J. Mol. Biol. 147, 195–197 (1981)

22. Socha, B.: Multithreaded execution of the Smith-Waterman algorithm in the query language for protein secondary structures. MSc thesis, supervised by Mrozek D., Silesian University of Technology, Gliwice, Poland (2013)

23. Stephens, S., Chen, J.Y., Thomas, Sh.: ODM BLAST: sequence homology search in the RDBMS. In: Bulletin of the IEEE Computer Society Technical Committee on Data Engineering (2004)

24. Tata, S., Patel, J.M., Friedman, J.S., Swaroop, A.: Declarative querying for biological sequences. In: Proceedings of 22nd International Conference on Data Engineering, IEEE Computer Society, 2006, pp. 87–98 (2006)

25. Wang, Y., Sunderraman, R., Tian, H.: A domain specific data management architecture for protein structure data. In: Proceedings of 28th IEEE EMBS Annual International Conference, New York City, USA, pp. 5751–5754 (2006)

26. Wieczorek, D., Małysiak-Mrozek, B., Kozielski, S., Mrozek, D.: A metod for matching sequences of protein secondary structures. J. Med. Info. Technol. **16**, 133–137 (2010)

27. Wieczorek, D., Małysiak-Mrozek, B., Kozielski, S., Mrozek, D.: A declarative query language for protein secondary structures. J. Med. Info. Technol. **16**, 139–148 (2010)
28. Yang, Y., Faraggi, E., Zhao, H., Zhou, Y.: Improving protein fold recognition and template-based modeling by employing probabilistic-based matching between predicted one-dimensional structural properties of the query and corresponding native properties of templates. Bioinformatics **27**, 2076–2082 (2011)
29. Zomaya, A.Y.: Parallel Computing for Bioinformatics and Computational Biology: Models, Enabling Technologies, and Case Studies, 1st edn. Wiley-Interscience, New York (2006)

Chapter 3
Parallel CUDA-Based Protein 3D Structure Similarity Searching

The structural alignment between two proteins: is there a unique answer?

Adam Godzik, 1996

Abstract Finding common molecular substructures in complex 3D protein structures is still challenging. This is especially visible when scanning entire databases containing tens or even hundreds of thousands protein structures. Graphics processing units (GPUs) and general purpose graphics processing units (GPG-PUs) promise to give a high speedup of many time-consuming and computationally demanding processes over their original implementations on CPUs. In this chapter, we will see that a massive parallelization of the 3D structure similarity searching on many core CUDA-enabled GPU devices leads to reduction of the execution time of the process and allows to perform it in real time.

Keywords Proteins · 3D protein structure · Tertiary structure · Similarity searching · Structure matching · Structure comparison · Structure alignment · Parallel computing · GPU · CUDA

3.1 Introduction

Protein 3D structure similarity searching is a process in which a given protein structure is compared to another protein structure or a set of protein structures collected in a database. The aim of the process is to find matching fragments of compared protein structures. On the basis of the similarities found during this process, scientists can draw useful conclusions about the common ancestry of the proteins, and thus the organisms (that the proteins came from), their evolutionary relationships, functional similarities, existence of common functional regions, and many other things [6]. This process is especially important in situations, where sequence similarity searches fail or deliver too few clues [15]. There are also other processes in which protein structure

D. Mrozek, *High-Performance Computational Solutions in Protein Bioinformatics*,
SpringerBriefs in Computer Science, DOI: 10.1007/978-3-319-06971-5_3,
© The Author(s) 2014

similarity searching plays a supportive role, such as in the validation of predicted protein models [23]. Finally, we believe that in the very near future, scientists will have the opportunity to study beautiful structures of proteins as a regular diagnostic procedure that will utilize comparison methods to highlight areas of proteins that are inadequately constructed, leading to dysfunctions of the body and serious diseases. This goal is currently motivating work leading to the development of similarity searching methods that return results in real time.

3.1.1 What Makes the Problem

Although protein structure similarity searching belongs to a group of the primary tasks performed in a structural bioinformatics, it is still a very difficult and time-consuming process. There are three key factors deciding on this:

1. the 3D structures of proteins are highly complex,
2. the similarity searching process is computationally complex,
3. the number of 3D structures stored in macromolecular data repositories such as the Protein Data Bank (PDB) [2] is growing exponentially.

Among these three problems, the bioinformaticians can attempt to easy the second one by developing new, more efficient algorithms, and to—at least partially—help with the first one by using appropriate representative features of protein 3D structures that can then be fed into their algorithms. The collection of algorithms that have been developed for protein structure similarity searching over the last two decades is large, and included methods such as VAST [13], DALI [17, 18], LOCK2 [46], FATCAT [52], CTSS [9], CE [47], FAST [56], and others [33, 42]. These methods use various representative features when performing protein structure similarity searches in order to reduce the huge search space. For example, local geometric features and selected biological characteristics are used in the CTSS [9] algorithm. Shape signatures that include information on C_α atom positions, torsional angles, and types of the secondary structure present are calculated for each residue in a protein structure. A very popular DALI algorithm [17, 18] compares proteins based on distance matrices built for each of the compared proteins. Each cell of a distance matrix contains the distance between the C_α atoms of every pair of residues in the same structure (inter-residue distances). Fragments of 6×6 elements of the matrix are called *contact patterns*, which are compared between two proteins to find the best match. On the other hand, the VAST algorithm [13], which is available through the web site of the National Center for Biotechnology Information (NCBI), uses secondary structure elements (SSEs: α-helices and β-sheets), which form the cores of the compared proteins. These SSEs are then mapped to the representative vectors, which simplifies the analysis and comparison process. During the comparison, the algorithm attempts to match vectors of pairs of protein structures. Other methods, like LOCK2 [46], also utilize the SSE representation of protein structures in the comparison process. The CE [47] algorithm uses the combinatorial extension of alignment path formed by aligned fragment pairs (AFPs). AFPs are fragments of both structures that indicate clear structural similarity and are described by local geometrical features, including positions of C_α atoms. The idea of AFPs is also used in the FATCAT [52]. A more detailed overview of methodologies used for protein structure comparison and similarity searching is given in [3, 7, 8].

Even though better methods are developed every year, performing a protein structure similarity search against a whole database of protein 3D structures is still a challenge. As it was shown in the works [27, 30] on the effectiveness and scalability of the process, performing the search with the FATCAT algorithm for a sample query protein structure using twenty alignment agents working in parallel took 25 h (without applying any additional acceleration techniques). Tests were carried out using a database containing 3D structures of 106,858 protein chains. This shows how time-consuming the process is, and it is one of the main motivations for designing and developing the new methods that are reported every year, such as RAPIDO [31], FS-EAST [32], DEDAL [11], MICAN [29], CASSERT [33], ClusCo [20], and others [36, 38, 53–55].

3.1.2 CUDA Architecture and Construction of GPU Devices

The evolution of computer science and computer architectures has led to (and will continue to lead to) new hardware solutions that can be used to accelerate various time-consuming processes. Recent years have shown that promising results can be obtained by using graphics processing units (GPUs) and general purpose graphics processing units (GPGPUs). GPU devices, which were originally conceived as a means to render increasingly complex computer graphics, can now be used to perform computations that are required in completely different domains. For this reason, GPU devices, especially those utilizing the NVidia Compute Unified Device Architecture (CUDA) [35, 43], are now widely used to solve computationally intensive problems, including those encountered in bioinformatics.

In GPU devices that support the CUDA architecture, high scalability is achieved by the hierarchical organization of *threads*, which are basic execution units. Threads execute, in parallel, user-defined procedures called *kernels*, which implement some computational logic working on different data. Each thread has its own index, the vector of the coordinates corresponding to its location in the one-, two-, or three-dimensional organizational structure called a *block*. Thread blocks form a one- or two-dimensional structure called a *grid*. Each thread block is processed by a streaming multiprocessor (SM), which has many scalar processor cores (SP). The number of multiprocessors and processor cores available depends on the type of GPU device. The GPU device has also two special function units, a multithreaded instruction unit (IU), a set of registers available for each thread block, and several types of memory (Fig. 3.1).

Threads can access global memory, which is the off-chip memory that has a relatively low bandwidth but provides a high storage capacity. Each thread also has access to the on-chip read/write shared memory as well as the read-only constant memory and texture memory, both of which are cached on-chip. Access to these

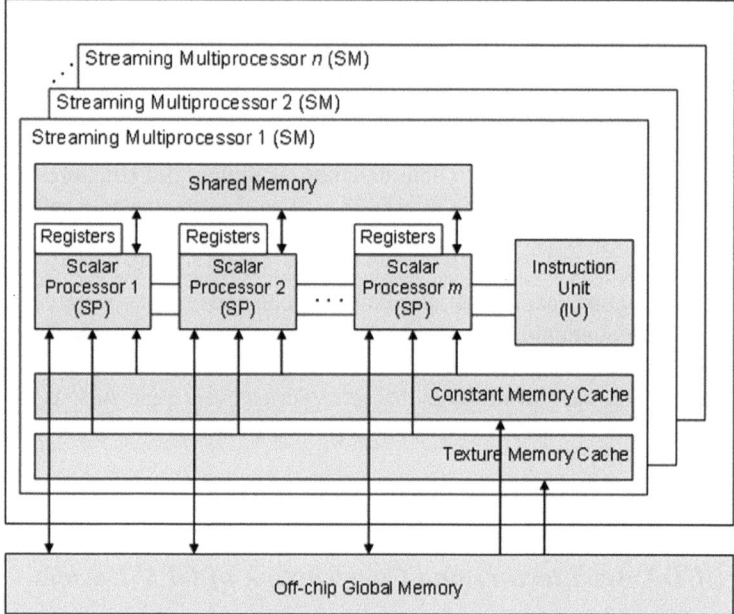

Fig. 3.1 Architecture of the GPU computing device, showing streaming multiprocessors, scalar processor cores, registers, and global, shared, constant, and texture memories

three types of memories is much faster than that to the global memory, but they all provide limited storage space and are used in specific situations.

Multiprocessors employ a new architecture, called single instruction, multiple thread (SIMT). In this architecture, a multiprocessor maps each thread to a scalar processor core, where each thread executes independently with its own instruction address and register state. The multiprocessor SIMT unit creates, manages, schedules, and executes threads in groups of 32 parallel threads called *warps*. Threads in the warp perform the same instructions, but operate on different data, as in the SIMD (single instruction, multiple data) architecture. Therefore, appropriate preparation and arrangement of data is highly desirable before the kernel execution begins, and this is one of the factors that influence the efficiency of any GPU-based implementation [35].

3.1.3 CUDA-Enabled GPUs in Processing Biological Data

The computational potential of GPU devices has been also noticed by specialists working in the domain of life sciences, including bioinformatics. Given the successful applications of GPUs in the field of sequence similarity [24–26, 28, 39, 44, 50], phylogenetics [51], molecular dynamics [12, 40], and microarray data analysis [5], it is clear that GPU devices are beginning to play a significant role in the 3D protein structure similarity searching.

It is worth mentioning two GPU-based implementations of the process. These methods use different representations of protein structures and different computational procedures, but demonstrate a clear improvement in performance over the CPU-based implementations. The first one, *SA Tableau Search* presented in [49], uses simulated annealing for tableau-based protein structure similarity searching. Tableaux are based on orientations of secondary structure elements and distance matrices. The GPU-based implementation of the algorithm parallelizes two areas: multiple iterations of the simulated annealing procedure and multiple comparisons of the query protein structure to many database structures. The second one, called *pssAlign* [37], consists of two alignment phases—*fragment-level alignment* and *residue-level alignment*. Both phases use dynamic programming [1]. In the *fragment-level alignment* phase so-called *seeds* between the target protein and each database protein are used to generate initial alignments. These seeds are represented by the locations of the C_α atoms. The initial alignments are then refined in the *residue-level alignment* phase. *pssAlign* parallelizes both alignment phases.

In the following sections, we will see the GPU-based implementation of the CASSERT [33], one of the newest algorithms for protein 3D structure similarity searching. Like *pssAlign*, CASSERT is based on two-phase alignment. However, it uses an extended set of structural features to describe protein structures, and the computational procedure differs too. Originally, CASSERT was designed and implemented as a CPU-based procedure, and its effectiveness is reported in [33]. Its GPU-based implementation will be referred as GPU-CASSERT throughout the chapter.

3.2 CASSERT for Protein Structures Similarity Searching

Three-dimensional protein structure similarity searching is typically realized by performing pairwise comparisons of the query protein (Q) specified by the user with successive proteins (D) from the database of protein structures. Here, we will see how protein structures are represented in both phases of the comparison process performed by the CASSERT.

Let us assume that Q represents the structure of the query protein that is q residues (amino acids) long, and D is the structure of a candidate protein in the database that is d residues (amino acids) long.

In the first phase of the alignment algorithm, protein structures Q and D are compared by aligning their *reduced chains of secondary structures* formed by secondary structure elements SE_i:

$$Q = (SE_1^Q, SE_2^Q, ..., SE_n^Q), \tag{3.1}$$

where $n \leq q$ is the number of secondary structures in the chain of the query protein Q, and

$$D = (SE_1^D, SE_2^D, ..., SE_m^D), \tag{3.2}$$

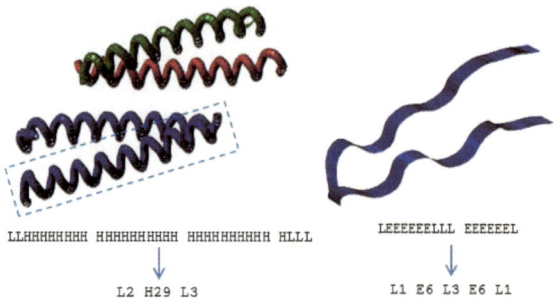

```
LLHHHHHHH HHHHHHHHH HHHHHHHHH HLLL          LEEEEEELLL EEEEEEL
            ↓                                    ↓
        L2 H29 L3                           L1 E6 L3 E6 L1
```

Fig. 3.2 Secondary structure elements: (*left*) four α-helices in a sample structure [PDBID: 1CE9], (*right*) two β-strands joined by a loop in a sample structure [PDB ID: 1E0Q]; visualized by MViewer [48]. Full and reduced chains of secondary structure elements for marked subunit (*left*) and the whole structure (*right*) are visible below

where $m \leq d$ is the number of secondary structures in the chain of the database protein D.

Each element SE$_i$, which is a part of the chain that has been selected on the basis of its secondary structure, is characterized by two values, i.e.,

$$SE_i = [SSE_i, L_i], \qquad (3.3)$$

where SSE$_i$ describes the type of the secondary structure selected, and L_i is the length of the ith element SE$_i$ (measured in residues). The alignment method distinguishes between three basic types of secondary structures (Fig. 3.2):

- α-helix (H),
- β-sheet or β-strand (E),
- loop, turn, coil, or undetermined structure (L).

Elements SE$_i^Q$ and SE$_j^D$, hereinafter referred to as SE regions or SE fragments, are built from groups of adjacent amino acids that form the same type of secondary structure. For example, six successive residues folded into an α-helix form one SE region. Hence, the overall protein structures are, at this stage, represented by the *reduced chains of secondary structures*.

In the second phase of the alignment algorithm, protein structures Q and D are represented in more detail. At the residue level, successive residues are described by so-called *molecular residue descriptors* s_i. Proteins are represented as chains of descriptors s_i:

$$Q = (s_1^Q, s_2^Q, ..., s_q^Q), \qquad (3.4)$$

where q is the length of the query protein Q (i.e., the number of residues it contains), and each s_i^Q corresponds to the ith residue in the chain of protein Q,

$$D = (s_1^D, s_2^D, ..., s_d^D), \qquad (3.5)$$

Fig. 3.3 Structural features included in molecular residue descriptors marked on part of a sample protein structure: residue type (Met, Gln, Ile, Phe), secondary structure type (β-strand in this case), length of the vector between C_α atoms ($|C_i|$) and the γ angle

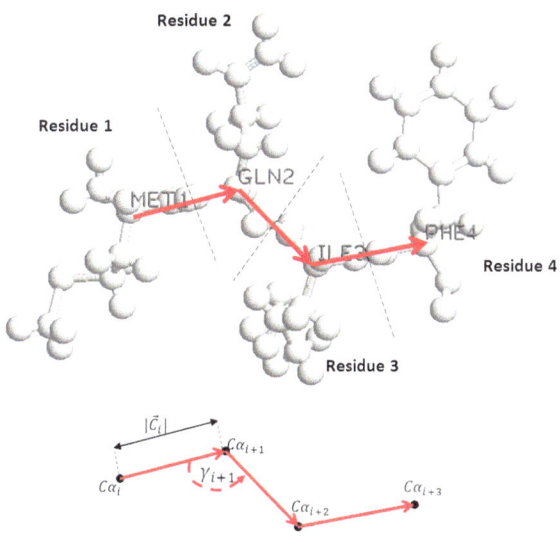

where d is the length of the database protein D, and each s_i^D corresponds to the ith residue in the chain of protein D.

Each descriptor s_i is defined by the following vector of features:

$$s_i = \langle |\mathbf{C_i}|, \gamma_i, \mathrm{SSE}_i, r_i \rangle, \qquad (3.6)$$

where $|\mathbf{C_i}|$ is the length of vector between C_α atoms of the ith and $(i+1)$th amino acid in a protein chain, γ_i is the angle between successive vectors $\mathbf{C_i}$ and $\mathbf{C_{i+1}}$, SSE_i is the type of secondary structure formed by the ith residue, r_i is a type of amino acid (Fig. 3.3).

3.2.1 General Course of the Matching Method

Pairwise comparisons of protein 3D structures are performed using the matching method, which consists of two phases (Fig. 3.4):

1. The first phase involves the coarse alignment of spatial structures represented by secondary structure elements (SSEs). This is the *low resolution alignment* phase, because groups of amino acids occurring in each structure are grouped into one representative element (the SE region). This phase allows us to run fast alignments in which small similarity matrices are constructed. This eliminates the need for computationally costly alignments for proteins that are entirely dissimilar. Proteins that exhibit secondary structures similarity are subjected to more thorough analysis in the second phase.

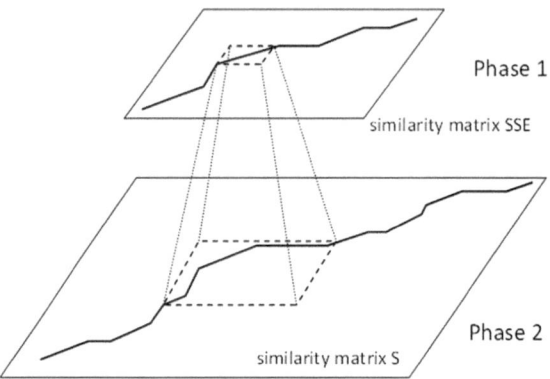

Fig. 3.4 Overview of the two-phase alignment algorithm. In phase 1, low-resolution alignment is performed; protein structures are represented as reduced chains of secondary structures; the similarity matrix SSE used in the alignment is small—proportional to the number of secondary structures in both proteins. In phase 2, high-resolution alignment is performed; protein structures are represented as chains of molecular residue descriptors; the similarity matrix S used in the alignment is therefore large—proportional to the length of both proteins

2. The second phase involves the detailed alignment of spatial structures represented by the molecular residue descriptors. This alignment is performed based on the results of the coarse alignment realized in the first phase. The second phase is the *high-resolution alignment* phase, because amino acids are not grouped in it. Instead, each amino acid found in the structure is represented by the corresponding molecular residue descriptor s_i. Therefore, in this phase CASSERT aligns sequences of molecular residue descriptors using much larger similarity matrices than were utilized in the first phase. In the second phase, the algorithm analyzes more features describing protein structures, and the protein itself is represented in more detail.

In both phases, the alignments are carried out using dynamic programming procedures that are specifically adapted to the molecular descriptions of protein structures in each phase. The detailed courses of both alignment phases are shown in the following sections.

3.2.2 First Phase: Low-Resolution Alignment

The low-resolution alignment phase is performed in order to filter out molecules that do not show secondary structural similarity. Originally, this phase was also used to establish initial alignments that were projected onto the similarity matrix in the second phase. However, since both phases are executed independently in the GPU-based implementation, alignment paths are not transferred between alignment phases in the GPU-based approach.

In order to match the structures of proteins Q and D that are represented as reduced chains of secondary structures, the algorithm builds the similarity matrix SSE of size $n \times m$, where n and m describe the number of secondary structures in the compared chains of proteins Q and D. Successive cells of the SSE matrix are filled according to the following rules:

For $0 \le i \le n$ and $0 \le j \le m$:

$$SSE_{i,0} = SSE_{0,j} = 0, \tag{3.7}$$

$$SSE_{i,j}^{(1)} = SSE_{i-1,j-1} + \delta_{ij}, \tag{3.8}$$

$$SSE_{i,j}^{(2)} = E_{i,j}, \tag{3.9}$$

$$SSE_{i,j}^{(3)} = F_{i,j}, \tag{3.10}$$

$$SSE_{i,j} = \max_{v=1..3} \{SSE_{i,j}^{(v)}, 0\}. \tag{3.11}$$

where δ_{ij} is the similarity reward, which reflects the degree of similarity between two regions SE_i^Q and SE_j^D of proteins Q and D, respectively, and vectors E and F define possible horizontal and vertical penalties for inserting a gap.

The similarity reward δ_{ij} takes values from the interval $[0, 1]$, where 0 means no similarity and 1 means that the regions are identical. The degree of similarity is calculated using the formula:

$$\delta_{ij} = \sigma_{ij} - \left(\sigma_{ij} * \frac{|L_j^D - L_i^Q|}{(L_j^D + L_i^Q)} \right), \tag{3.12}$$

where L_i^Q, L_j^D are lengths of compared regions SE_i^Q and SE_j^D, while σ_{ij} describes the similarity degree of secondary structures building ith and jth SE regions of compared proteins Q and D. This parameter can take three possible values according to the following rules:

(i) $\sigma_{ij} = 1$, when both SE regions have the same secondary structure of α-helix or β-strand;

(ii) $\sigma_{ij} = 0.5$, when at least one of the regions is a loop, turn, coil, or its secondary structure is undefined;

(iii) $\sigma_{ij} = 0$, when one of the regions has the construction of α-helix and the second the construction of β-strand.

Values of gap penalty vectors are calculated as follows:

$$E_{i,j} = \max \begin{cases} E_{i-1,j} - g_E \\ SSE_{i-1,j} - g_O \end{cases}, \tag{3.13}$$

$$F_{i,j} = \max \begin{cases} F_{i,j-1} - g_E \\ SSE_{i,j-1} - g_O \end{cases}. \tag{3.14}$$

In order to assess the similarity between two reduced chains of secondary structures, CASSERT uses the Score measure, which is equal to the highest value in the similarity matrix SSE:

$$\text{Score} = \max \{\text{SSE}_{i,j}\}. \tag{3.15}$$

Auxiliary vectors E and F allow us to perform alignment procedure and to calculate the Score similarity measure in linear space, because the value of cell $\text{SSE}_{i,j}$ depends only on the value of cell $\text{SSE}_{i-1,j-1}$, $\text{SSE}_{i-1,j}$, and $\text{SSE}_{i,j-1}$. During the calculation of the similarity matrix SSE CASSERT has to store the position of the maximum value of the Score in the matrix as well as the value itself.

3.2.3 Second Phase: High-Resolution Alignment

Molecules that pass the first phase (based on the user-defined cutoff value) are further aligned in the second phase. A database protein structure is qualified to the second phase if the following condition is satisfied:

$$\frac{\text{Score}^{QD}}{\text{Score}^{QQ}} \geq Q_t, \tag{3.16}$$

where Score^{QD} is a similarity measure employed when matching the query protein structure to the database protein structure, Score^{QQ} is the similarity measure obtained when matching the query protein structure to itself (i.e., the maximum Score that the compared chain can achieve), and $Q_t \in [0, 1]$ is a user defined qualification threshold for structural similarity.

The second phase is carried out similarly to the first phase, except that the alignment is carried out on the residue level, where aligned molecules Q and D are represented by chains of molecular residue descriptors. However, the way that GPU-CASSERT calculates the similarity reward for the two compared residue molecular descriptors s_i and s_j is different. The similarity reward ss_{ij} is calculated according to the following formula:

$$ss_{ij} = w_C \sigma_{ij}^C + w_\gamma \sigma_{ij}^\gamma + w_{\text{SSE}} \sigma_{ij}^{\text{SSE}} + w_r \sigma_{ij}^r, \tag{3.17}$$

where σ_{ij}^C is the degree of similarity of a pair of vectors $\mathbf{C}_i^Q$ and $\mathbf{C}_j^D$ in proteins Q and D, σ_{ij}^γ is the similarity of angles γ_i^Q and γ_j^D in proteins Q and D, σ_{ij}^{SSE} is the degree of similarity of secondary structures of residues i and j (calculated according to the rules (i)–(iii) listed for the first phase), σ_{ij}^r is the degree of similarity of residues defined by means of the BLOSUM62 substitution matrix [16] normalized to range of [0, 1], and w_C, w_γ, w_{SSE}, w_r are the weights of all of the components (with default value of 1).

The similarity of vectors C_i^Q and C_j^D is defined according to the formula:

$$\sigma_{ij}^C = e^{-\left(|C_i^Q| - |C_j^D|\right)^2},$$ (3.18)

where $|C_i^Q|$ and $|C_j^D|$ are the lengths of vectors C_i^Q and C_j^D, respectively, and the similarity of the angles γ_i^Q and γ_j^D is defined as follows:

$$\sigma_{ij}^\gamma = e^{-\left(\gamma_i^Q - \gamma_j^D\right)^2}.$$ (3.19)

In high-resolution alignment, the value of the degree of similarity of molecular residue descriptors ss_{ij} (Eq. 3.17) replaces the similarity reward δ_{ij} (Eq. 3.8).

The relative strength of each component in the similarity search (Eq. 3.17) can be controlled using participation weights. The default values for each is 1, but this can be changed by the user. For example, researchers who are looking for surprising structural similarities but no sequence similarity can disable the component for the primary structure by setting the value of $w_r = 0$.

The Score similarity measure, the basic measure of the similarity of protein structures, is also calculated in this phase. Its value incorporates all possible rewards for a match, mismatch penalties, and penalties for inserting gaps in the alignment. The Score is also used to rank highly similar proteins that are returned by the GPU-CASSERT.

3.2.4 Third Phase: Structural Superposition and Alignment Visualization

In the third phase, the algorithm performs superposition of protein structures on the basis of aligned chains of molecular residue descriptors. The purpose of this step is to match two protein structures by performing a set of rotation and translation operations that minimizes the root mean square deviation (RMSD):

$$\text{RMSD} = \sqrt{\frac{1}{N} \sum_{i=1}^{N} d_i^2},$$ (3.20)

where N is the number of aligned C_α atoms in the protein backbones, and d_i is the distance between the ith pair of atoms.

Two approaches are widely used to complete this step. One of the approaches uses quaternions [19]. CASSERT uses the approach proposed by Kabsch [21, 22] that makes use of the Singular Value Decomposition (SVD) technique. These two approaches are said to be computationally equivalent [10], but there can be some circumstances deciding that one can be more convenient than the other.

CASSERT performs the superposition of protein structures on the CPU of the host workstation. In this phase, CASSERT also calculates the full similarity matrix S in order to allow backtracking from the maximum value and full visualization of the structural alignment at the residue level. This step is performed on the CPU of the host and only for a limited number (M, which is configured by the user) of the most similar molecules.

3.3 GPU-Based Implementation of the CASSERT

GPU devices can accelerate calculation speeds greatly, but this also requires the application of an appropriate programming model. In terms of 3D protein structure similarity searching, this also involves preparation of data that will be processed. In this subsection, we will focus on the GPU-based implementation of the CASSERT, data preparation, and we will gain insight into the implementation details of both alignment phases. A part of the work was carried out by Brożek [4], my associate in this project.

3.3.1 Data Preparation

Early tests of the first implementations of the CASSERT algorithm on GPU devices showed that read operations from the database system storing structural data were too slow. Therefore, the present implementation of the GPU-CASSERT does not read data directly from the database, because single execution of the searching procedure would take too long. GPU-CASSERT uses binary files instead. These files contain data packages that are ready to be sent to the GPU device. The only data that are read directly from the database are those that describe the query protein structure Q. But, even in this situation, the data are stored in an appropriate way in binary files. Using binary files with data packages allows the initialization time of the GPU device to be reduced severalfold. This is necessary to ensure that the GPU-CASSERT has a fast response time.

Binary files are refreshed in two cases:

- changes in the content of a database,
- changes in parameters affecting the construction of data packages.

Data packages that are sent to the GPU device have the same general structure, regardless of what is stored inside.

Due to the size of the data packages utilized by the CASSERT algorithm, these packages are placed in the global memory of the GPU device. As we know from the Sect. 3.1.2 when discussed GPUs and the CUDA, global memory is the slowest type of memory available. For this reason, it is worth minimizing the number of accesses made of this type of memory.

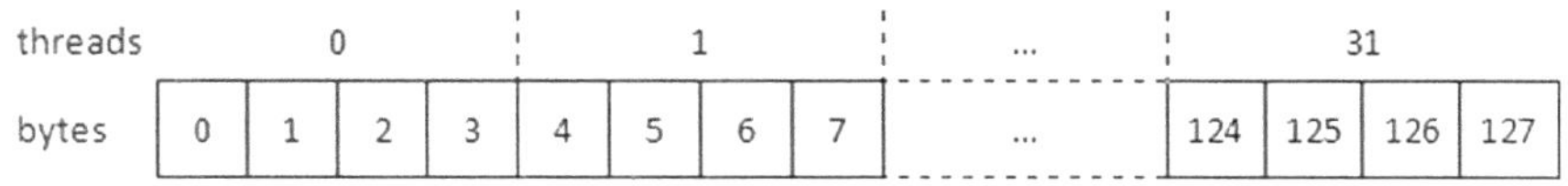

threads		0				1				...		31			
bytes	0	1	2	3	4	5	6	7		...		124	125	126	127

Fig. 3.5 Preferred allocation of 128-byte memory segment to warp threads. Thread 0 takes first 4 bytes of the transaction, thread 1 takes the next 4 bytes, etc

Access operations are carried out in 32-, 64-, or 128-byte transactions. When the warp (which is composed of 32 threads) reaches the read/write operation, the GPU device attempts to perform this operation using a minimum number of transactions. Basically, the greater the number of transactions needed, the greater the amount of unnecessary data transmitted. This unnecessary overhead can be minimized for CUDA 2.x if memory cells that are read by all warp threads are located within a single 128-byte memory segment. In order to satisfy this condition, the address of this area must be aligned to 128 bytes and the threads need to read data from adjacent memory cells. For devices with compute capabilities of 1.0 or 1.1, upon which the GPU-CASSERT can also run, there is the additional restriction that warp threads must be in the same order as memory cells being read [35]. If these conditions are met, we can get 4 bytes of data for each of the threads in a single 128-byte transaction. These 4 bytes correspond to a single number of the type *int* or *float*, which is used while encoding data in data packages. The preferred allocation of 128-byte memory segment to threads is presented in Fig. 3.5.

Data are transmitted to the GPU device in the form of a two-dimensional array of unsigned integers (Fig. 3.6). The array is organized in row-major order. This means that the cells in adjacent columns are located next to each other in the memory. This has an important influence on performance when processing an array, because contiguous array cells can usually be accessed more quickly than cells that are not contiguous. Each column of the array is assigned to a single block thread. Threads start at an index given by the following code:

```
int tid=blockIdx.x*blockDim.x+threadIdx.x;
```

where `blockIdx.x` is the block index along the x dimension (GPU-CASSERT uses a one-dimensional blocks), `blockDim.x` stores the number of threads along the x dimension of the block, and `threadIdx.x` is the thread index within the block.

A single chain of structural descriptors is stored in a single column of the array (Fig. 3.6). Such a solution satisfies the condition that contiguous addresses must be read, because block threads will always read adjacent cells, moving from the beginning to the end of the chain (from top to bottom). Every cell in the array is 4 bytes in size, so the transfer of data to a wrap's 32 threads will be made in one 128-byte read transaction. This allows to take a full advantage of data transfer from the memory to the registers of the GPU device. This way of organizing data in memory is used and described in [28, 39].

Another factor affecting the performance is the density at which the data are packed in memory cells. The distribution of data in memory cells depends on the

threads	0	1	2	3	4	5	6	7	...	31
	S_1	S_1	S_1	S_1	S_1	S_1	S_1	S_1	...	S_1
	S_2	S_2	S_2	S_2	S_2	S_2	S_2	S_2	...	S_2
	S_3	S_3	S_3	S_3	S_3	S_3	S_3	S_3	...	S_3
	S_4	S_4	S_4	S_4	S_4	S_4	S_4	S_4	...	S_4
	S_5	S_5	S_5	S_5	S_5	S_5	S_5	S_5	...	S_5
	S_6	S_6	S_6	S_6	S_6	S_6	S_6	S_6	...	S_6
	S_7	S_7	S_7	S_7	S_7	S_7	S_7	S_7	...	S_7
	...	...	...	...	...	...	...	...	...	...

Fig. 3.6 Arrangement of chains of structural descriptors S_1, S_2, S_3, ... in a memory array. Block threads are assigned to particular columns. One column stores one chain of structural descriptors. Each cell contains 4 bytes of data (structural descriptors). All *block threads* read contiguous memory areas (coalesced access)

phase of the algorithm and the type of structural descriptors that are used in the phase. There are five types of data that are sent to the memory of the GPU device:

- reduced chains of secondary structures formed by secondary structure elements SE_i (phase 1),
- secondary structure elements SSE_i that are components of nonreduced chains of molecular residue descriptors (phase 2),
- amino acid residue types r_i that are components of nonreduced chains of molecular residue descriptors (phase 2),
- lengths of the vectors between C_α atoms of subsequent residues that are components of nonreduced chains of molecular residue descriptors (phase 2),
- γ_i angles between successive vectors $\mathbf{C_i}$ and $\mathbf{C_{i+1}}$ that are components of nonreduced chains of molecular residue descriptors (phase 2).

Regardless of the type of data present in the memory cells, the chains included in the package may be of various lengths. For this reason, all chains of structural descriptors are aligned to the length of the longest chain. Empty cells are filled with zeros. In principle, comparing these zeros in the course of the algorithm does not affect the scoring system assumed and the final results.

Chains of structural descriptors contained in a data package are sorted by their lengths in ascending order. In this way, we minimize differences in processing time for individual block threads and their idle times (threads that have already completed their work must wait for the other threads to finish processing). A similar method is used in the work presented in [28, 39].

Data packages are divided into subpackages. Each subpackage consists of 32 chains of structural descriptors. This is exactly the same as the number of warp threads.

3.3.2 Implementation of Two-Phase Structural Alignment in a GPU

Implementation of the two-phase structural alignment algorithm in a GPU with the CUDA requires a dedicated approach. GPU-CASSERT operates according to the Algorithm 1.

In both alignment phases, the similarity matrix is stored in the global memory of the GPU device as an array of the type *float*. This means that a read/write of a single element requires just one transaction. It is also worth noting that, due to memory restrictions, each thread remembers only the last row of the similarity matrix. This is sufficient to determine the maximum element of the similarity matrix, which also provides a value for the Score similarity measure, which is needed to check whether a database structure qualifies for the second phase. The similarity measure alone is sufficient to assess the quality of the alignment before the second phase.

Algorithm 1 GPU-CASSERT: a general algorithm
___
1: Read data packages describing database protein structures from binary files
2: Read query protein structure (Q) from database and create appropriate data package with query profile
3: **for all** database proteins D **do**
4:     Perform (in parallel) the first phase of the structural alignment on the GPU device
5: **end for**
6: Qualify proteins for the second phase according to formula 3.16
7: Prepare data packages describing database protein structures for the second phase
8: Read data packages describing database protein structures from binary files
9: Read query protein structure (Q) from database and create appropriate data packages with query profiles
10: **for all** qualified database proteins D **do**
11:     Perform (in parallel) the second phase of the structural alignment on the GPU device
12: **end for**
13: Return a list of the top M database molecules that are most similar to the query molecule, together with similarity measures
14: **if** the user wants to visualize the alignment **then**
15:     Perform the second phase on the CPU of the host computer for molecules from the list of the most similar ones to the query molecule returned by the GPU device
16:     Perform structural superposition
17:     Return alignment visualization to the user
18: **end if**
___

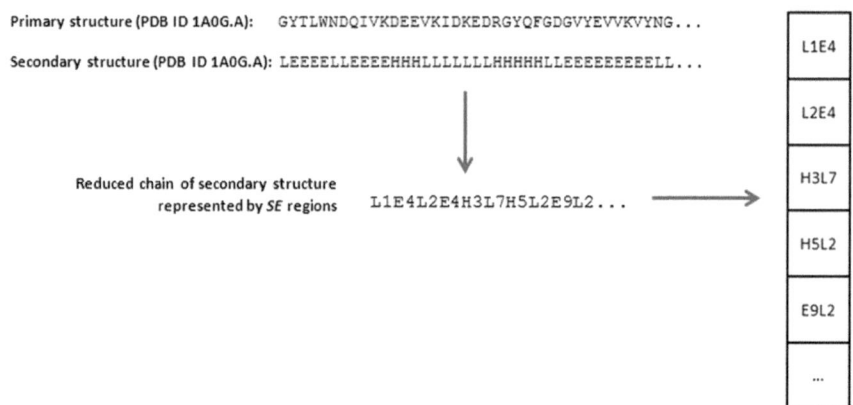

Fig. 3.7 Encoding a reduced chain of secondary structures in a data package. The secondary structure of the protein is first translated to a reduced chain of SE regions. Subsequently, every two SE regions are placed in a data package in the manner shown, taking up 4 bytes, and in such a way they are loaded to the global memory of the GPU device

On the other hand, the second phase is performed on the GPU device for all qualified structures, and once again on the CPU of the host for the database proteins that are most similar to the query molecule in order to get alignment paths and to perform structural superposition. As a result, the user obtains a list of the structures that match most closely to the query structure and a visualization of the local alignments of these structures at the residue level.

3.3.3 First Phase of Structural Alignment in the GPU

The first phase requires data to be delivered in the form of data packages containing reduced chains of secondary structures (SE regions). Separate data packages are built for the query protein and candidate protein structures from the database. For the purpose of processing, SE regions are encoded using two bytes: one byte for the type of secondary structure and one byte for its length (Eq. 3.3). Types of secondary structures are mapped to integers. In Sect. 3.3.1, where we talked about the overall structure of a data package, we also mentioned that the data in memory are arranged into 4-byte cells. In such a 4-byte cell, we can store two encoded SE regions. This is illustrated in Fig. 3.7.

The data package for the query chain of secondary structures is built on the basis of a slightly different principle. If it was created in the same way as the data packages for database structures, then in order to extract the similarity coefficient of secondary structures $\sigma_{i,j}$ we would have to read the cell (SSE_i^A, SSE_j^B) from a predefined matrix of coefficients (a kind of substitution matrix constructed based on rules (i)–(iii) in the Sect. 3.2.2), which would affect performance negatively. We can avoid this by

Fig. 3.8 Encoding the reduced chain of secondary structure for query protein Q (*left*) and construction of the query profile (*right*). The query profile shows all possible (encoded) scores when comparing the reduced query chain of secondary structure to SE regions from candidate protein structures from the database

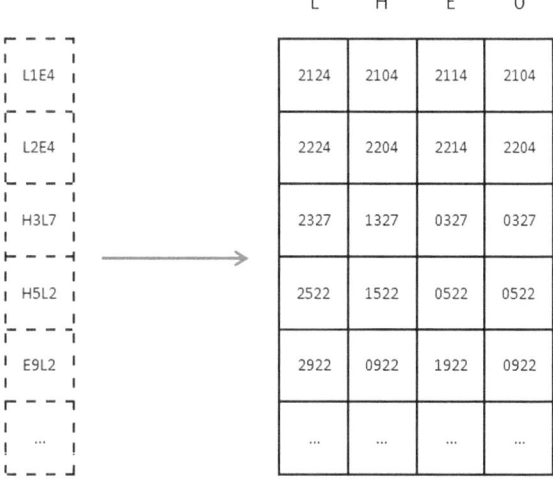

precomputing and writing all possible similarity coefficients directly into the data package of the query protein, creating something like the query-specific substitution matrix proposed in [41] and called a *query profile* in the GPU-based alignment algorithm for sequence similarity presented in [28]. Therefore, the data package for the query protein passes through an additional preparation step. For each SE region, four versions of the similarity coefficient are created, one for each of the secondary structure types and one for the neutral element 0 (as shown in Fig. 3.8). In the query profile created, the row index is defined by the index of structural region SE divided by 2, and the column index is defined by the type of secondary structure present (with the additional neutral element 0). The coefficients are converted to integers in order to fit them into 1 byte, according to the following rules:

- if coefficient $\sigma_{i,j} = 0$, it is encoded as 0,
- if coefficient $\sigma_{i,j} = 1$, it is encoded as 1,
- if coefficient $\sigma_{i,j} = 0.5$, it is encoded as 2.

Lengths of SE regions do not change. This process is illustrated in Fig. 3.8.

Once the data packages are loaded into the host memory and a data package for the reduced query chain is created, the program transfers data to the GPU device. To do this, it uses four streams. Each stream has its own memory buffers on the GPU device side and in the page-locked memory on the host side. The host loads data into the page-locked memory and then initiates an asynchronous data transfer to GPU device for each of the streams. This allows transmission to take place in parallel with the ongoing calculations, again improving performance. Results are received prior to the transfer of the next data package or after all available packages have been processed.

Block threads perform parallel alignments of reduced chains of secondary structures. Each block thread performs a pairwise alignment of the query protein versus

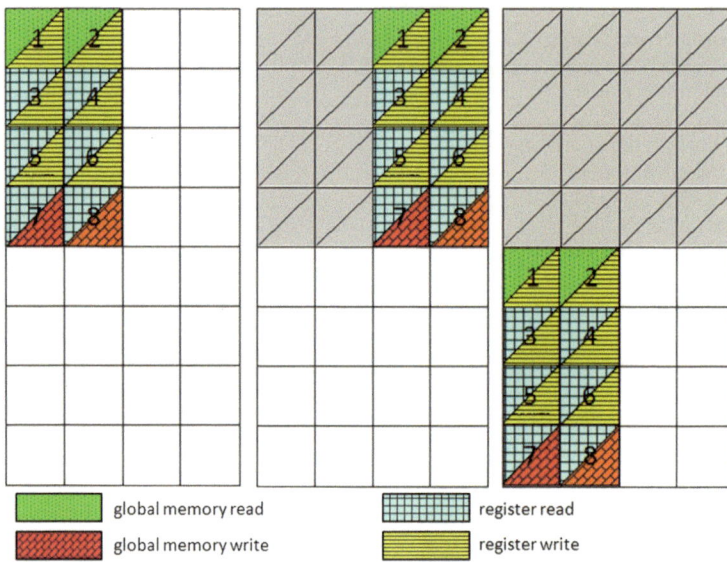

	global memory read		register read
	global memory write		register write

Fig. 3.9 Calculation of similarity matrix SSE. Structural elements (SE regions) of the candidate database structure are (virtually) located along the vertical edge of the matrix and SE regions of the query protein structure along the horizontal edge of the matrix. Calculations are performed in areas 2×4 in size. Values of the cells in these areas are calculated according to the given order. *Colors* reflect the type of read/write operation required and the memory resources that are affected

one candidate database protein. In order to limit the number of accesses to the global memory of the GPU device, the similarity matrix SSE is not calculated cell-by-cell but is divided into rectangular *areas* 2×4. Calculations are performed area-by-area, and row by row in each area, from left to right, as shown in Fig. 3.9.

Structural elements (SE regions) of the candidate database structure are (virtually) located along the vertical edge of the matrix, and SE regions of the query protein structure are located along the horizontal edge of the matrix. The pseudocode of the CUDA kernel for the calculation of the matrix SSE by a block thread is presented in Algorithm 2. The thread reads consecutive four elements SE_j^D, SE_{j+1}^D, SE_{j+2}^D, SE_{j+3}^D of the database protein from the global memory of the GPU and saves them in registers (lines 1–3). They will be used many times while calculating successive areas to the right of the leftmost area (Fig. 3.9, left and middle). Then, for each successive pairs of elements SE_i^Q, SE_{i+1}^Q of the query protein the thread reads values of $SSE_{i,j-1}$, $SSE_{i+1,j-1}$ and $F_{i,j-1}$, $F_{i+1,j-1}$ (calculated for the previous area, if any) from the global memory of the GPU and saves them in registers (lines 4–6). These values stored in registers will be swapped many times by current values of $SSE_{i,j}$, $SSE_{i+1,j}$ and $F_{i,j}$, $F_{i+1,j}$ during the calculation of area rows, since actually, at the end of the calculation we do not need the whole similarity matrix SSE, but the $Score^{QD}$ value. In the next step, for each row of the area the tread reads elements SE_i^Q, SE_{i+1}^Q of the query protein from the texture memory (lines 7–8). These two elements of the query protein correspond to only one row of the query profile. In line 9, the thread calculates values of $F_{i,j}$, $F_{i+1,j}$, $E_{i,j}$, $E_{i+1,j}$ and saves them in registers. They are required to calculate values of $SSE_{i,j}$ and $SSE_{i+1,j}$ of the matrix SSE according to formulas 3.7–3.11 (line 10). The value of $SSE_{i-1,j-1}$, which is also required for the calculation is stored in registers, as well. The values of $SSE_{i-1,j}$ and $E_{i-1,j}$ are equal to 0 for the leftmost areas (Fig. 3.9, left and right) or stored in registers after the calculation of the previous area (Fig. 3.9, middle).

Algorithm 2 Phase 1: kernel pseudocode for the calculation of the matrix SSE by a block thread (GM—global memory, TM—texture memory)

1: **for each** consecutive four elements SE_j^D, SE_{j+1}^D, SE_{j+2}^D, $SE_{j+3}^D : j = 1, ..., m$ **do**
2:     Reset registers
3:     Read from GM elements SE_j^D, ..., SE_{j+3}^D and save in registers
4:     **for each** successive pairs of elements $(SE_i^Q, SE_{i+1}^Q) : i = 1, ..., n$ **do**
5:         Read from GM values $SSE_{i,j-1}$, $SSE_{i+1,j-1}$ and save in registers
6:         Read from GM values $F_{i,j-1}$, $F_{i+1,j-1}$ and save in registers
7:         **for each** row of the area **do**
8:             Read from TM the element of the query profile that corresponds to (SE_i^Q, SE_{i+1}^Q)
9:             Calculate $F_{i,j}$, $F_{i+1,j}$, $E_{i,j}$, $E_{i+1,j}$ and save in registers
10:            Calculate $SSE_{i,j}$ and $SSE_{i+1,j}$ according to formulas 3.7- 3.11
11:            $Score^{QD} \leftarrow max(Score^{QD}, SSE_{i,j}, SSE_{i+1,j})$
12:            Save in registers values of $SSE_{i,j}$, $SSE_{i+1,j}$ for the next row of the area
13:            Save in register value of $S_{i+1,j}$ for the next pair (SE_i^Q, SE_{i+1}^Q) (next area)
14:         **end for**
15:         Save values of $SSE_{i,j}$, $SSE_{i+1,j}$ in the GM
16:         Save values of $F_{i,j}$, $F_{i+1,j}$ in the GM
17:         Save in register value of $SSE_{i+1,j}$ that will be used as diagonal value for another area
18:     **end for**
19: **end for**
20: Save in GM the value of $Score^{QD}/Score^{QQ}$

In line 11, a temporary value of the $Score^{QD}$ similarity measure is calculated. In line 12, current values of $SSE_{i,j}$, $SSE_{i+1,j}$ are stored in registers, replacing old values $SSE_{i,j-1}$, $SSE_{i+1,j-1}$. Values of $SSE_{i+1,j}$ for successive rows are also stored in additional set of registers for the calculation of the next area to the right (line 13). They serve as values $SSE_{i-1,j}$ for successive rows of the next area to the right of the current area. At the end of the calculation of the area, the thread writes values of $SSE_{i,j}$, $SSE_{i+1,j}$ and $F_{i,j}$, $F_{i+1,j}$, calculated for the last row, to the global memory (lines 15–16). They will be read and used again, when the thread processes the area below the current area. The value of $SSE_{i+1,j}$ for the last row of the area is stored in additional register (line 17). It will be used as diagonal value of $SSE_{i-1,j-1}$ at the beginning of the calculation of another area (down-right). Finally, when all cells of the matrix SSE are calculated, the thread knows the final $Score^{QD}$ and is able to calculate the value of $Score^{QD}/Score^{QQ}$, which will decide if the candidate protein is qualified for the second phase. The value is stored in the global memory (line 20).

During the calculation of each 2×4 area, the values of the four elements of the vector E representing the horizontal gap penalty and four elements of the matrix SSE to the left of the current area are stored in GPU registers. Four consecutive elements of the reduced chain of secondary structures for the database protein are read from the global memory once, before the calculation of each leftmost area of the matrix begins. They are also stored in GPU registers and reused during the calculation of other areas located on the right of the leftmost area. Calculation of a 2×4 area requires two reads and two writes to the global memory for the vector F representing the vertical gap penalty, and two reads and two writes for the similarity matrix SSE. It also requires four reads for the query profile placed in the texture memory. In total, the calculation of 8 cells of an *area* of the similarity matrix SSE requires eight read/write transactions to the global memory of the GPU device and four reads from the texture memory. The order of calculation of cells and read/write operations performed are shown in Fig. 3.9.

For the latest source codes of the GPU-CASSERT, please visit the project web site: http://zti.polsl.pl/dmrozek/science/gpucassert/cassert.htm

3.3.4 Second Phase of Structural Alignment in the GPU

After filtering candidate database proteins based on the qualification threshold Q_t, the program creates new, smaller data packages that are needed in the second phase. Separate data packages are built for each of the features included in the molecular residue descriptors. In data packages for amino acid types and secondary structure types, we can store elements for four successive molecular residue descriptors in every 4 bytes (and then in every 4-byte memory cell). The arrangement of bytes and cells in memory is similar to that used in the first phase (see Fig. 3.6). Vector lengths and angles occupy 4 bytes each, which is one cell of the prepared array in memory.

For the query protein structure, data packages for amino acid types and secondary structures are generated in a similar manner to how this is done in the first phase. The program creates separate query profiles for secondary structures and for residue types. The query profile for secondary structures is formed from the secondary structure similarity coefficients $\sigma_{i,j}$ in such a way that the row index is the index of the current element from the query chain divided by 4, and the column index is the type of the secondary structure of the element from the compared database protein (Fig. 3.10). The query profile for residue types is derived from the normalized BLOSUM62 substitution matrix in such a way that the row index is the index of the current element from the query chain divided by 4, and the column index is the type of the residue from the compared database chain. Data packages containing vector lengths and angles between these vectors, for the query protein structure, are created by rewriting these values to separate packages.

Transfer of data packages to the device is performed in the same manner as in the first phase. Four streams are used for this purpose. After the first part of data has been transferred to the GPU device, the high-resolution alignment procedure is

Fig. 3.10 Encoding the secondary structure elements (SSEs) from chains of molecular residue descriptors for query protein Q (*left*) and construction of the query profile (*right*). The query profile shows all possible (encoded) scores when comparing the query chain of SSEs to SSEs from candidate protein structures from the database

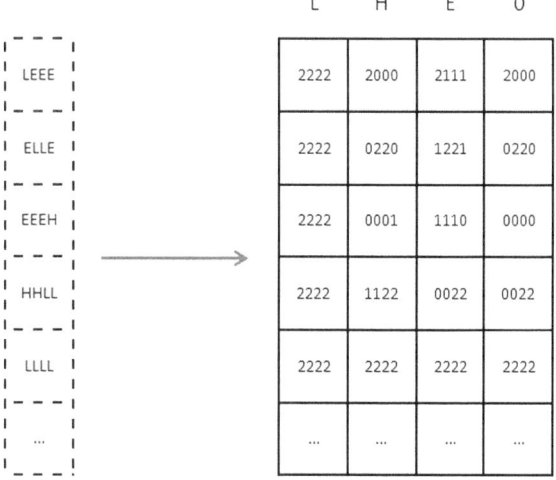

	L	H	E	0
LEEE	2222	2000	2111	2000
ELLE	2222	0220	1221	0220
EEEH	2222	0001	1110	0000
HHLL	2222	1122	0022	0022
LLLL	2222	2222	2222	2222
...	...	...	...	...

initiated. Block threads perform parallel alignments of chains of molecular residue descriptors. Each block thread performs a pairwise alignment of the query protein versus one candidate database protein. In order to limit the number of accesses to the global memory of the GPU device, the similarity matrix S is divided into rectangular *areas* of size 4×4. Calculations are performed area-by-area, and row-by-row inside areas, from left to right, as shown in Fig. 3.11.

Molecular residue descriptors of the candidate database structure are (virtually) located along the left vertical edge of the matrix S, and molecular residue descriptors of the query protein structure are located along the top horizontal edge of the matrix. During the calculation of each 4×4 area, the values of four elements of the vector E representing the horizontal gap penalty and the molecular residue descriptors for four successive elements of the database chain are stored in GPU registers. Calculation of a 4×4 area requires four reads and four writes to the global memory for the vector F representing the vertical gap penalty, and four reads and four writes for the similarity matrix S. It is also necessary to perform four reads for the query profile for secondary structures, four reads for the query profile for residue types, four reads for vector lengths, and four reads for angles between vectors. These reads are performed from the texture memory, where these structural features are placed and arranged in an appropriate manner. In total, the calculation of the 16 cells in each *area* of the similarity matrix S requires 16 read/write transactions to the global memory of the GPU device and 16 reads from the texture memory. The order of calculation of cells and the read/write operations performed are shown in Fig. 3.11. The kernel pseudocode is similar to the one presented for the first phase, with the exception that the thread processes 4×4 areas, which implies more I/O operations, and the similarity is calculated according to formula 3.17.

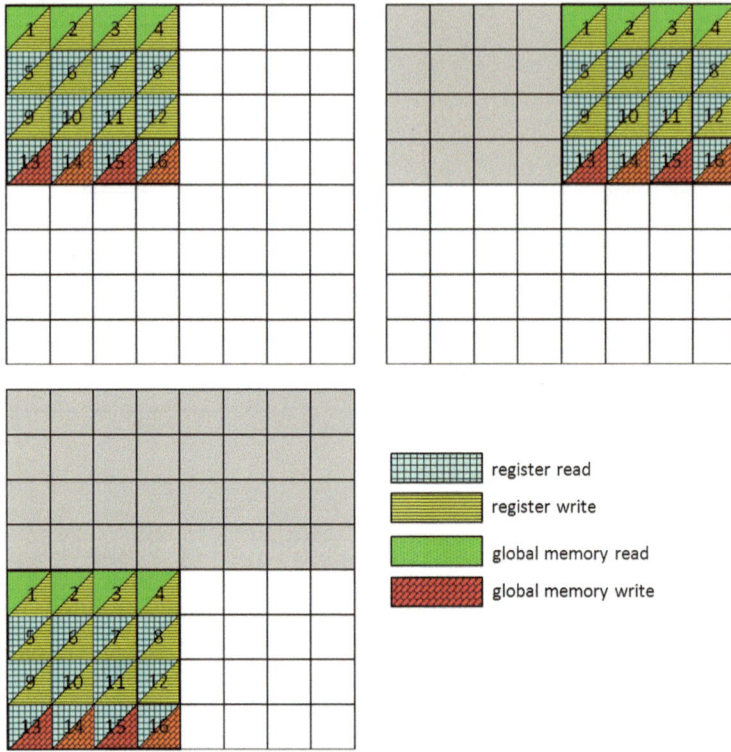

Fig. 3.11 Calculation of the similarity matrix S in the second phase of alignment. Molecular residue descriptors of the candidate database structure are (virtually) located along the vertical edge of the matrix and molecular residue descriptors of the query protein structure are located along the horizontal edge of the matrix. Calculations are performed in areas of size 4×4. Values of the cells in these areas are calculated according to the given order. *Colors* reflect the type of read/write operation that are required and the memory resources that are affected

3.4 GPU-CASSERT Efficiency Tests

The efficiency of the GPU-CASSERT algorithm was tested in a series of experiments. In this subsection, we will see results of these tests and we will compare the GPU-CASSERT to its CPU-based implementation that was published in [33]. Both implementations, i.e., the GPU-based and the CPU-based implementations, were tested on a Lenovo ThinkStation D20 with two Intel Xeon CPU E5620 2.4 GHz processors, 16 GB of RAM, and a GeForce GTX 560 Ti graphics card with 2 GB of GDDR5 memory. The workstation had the Microsoft Windows Server 2008 R2 Datacenter 64-bit operating system installed, together with the CUDA SDK version 4.2. The CUDA compute capability supported by the graphics card was 2.1. The graphics card had the following features:

Table 3.1 Query protein structures used in the performance tests

PDB ID	Chain	Length	PDB ID	Chain	Length
2CCE	A	29	1AYE	_	400
2A2B	A	40	2EPO	B	600
1BE3	G	80	1KK7	A	802
1A1A	B	101	1URJ	A	1027
1AYY	B	142	2PDA	A	1230
2RAS	A	199	2R93	A	1421
1TA3	B	300	2PFF	B	2005

- 8 multiprocessors (384 processing cores),
- 48 KB of shared memory per block,
- 64 KB of total constant memory,
- 32,768 registers per block,
- 2 GB of total global memory.

Tests were conducted using the DALI database (the same as that used by the DALI algorithm [17, 18]), which contained the structures for 105,580 protein chains. While testing performance, 14 selected query protein structures with lengths between 29 and 2005 amino acids were used. These were randomly selected molecules that represent different classes according to SCOP classification [34], i.e., all α, all β, $\alpha + \beta$, α/β, $\alpha\&\beta$, coiled coil proteins, and others. The list of query protein structures used in the tests performed in the present work is shown in Table 3.1.

Tests were performed using different qualification thresholds $Q_T = 0.01$, 0.2, 0.4, 0.6, 0.8 that the structures had to attain for them to pass from the first phase to the second phase of CASSERT. CASSERT execution times for $Q_T = 0.01$ and $Q_T = 0.2$ are shown in Fig. 3.12. The thresholds used were not chosen randomly. The $Q_T = 0.2$ is an experimentally determined threshold that filters out a reasonable number of structures based on the secondary structure similarity but still allows short local similarities to be found. This will be discussed further later in the sections. The $Q_T = 0.01$ means that almost no filtering is done based on the secondary structure similarity, and almost all structures in the database qualify for the second phase.

The results of the efficiency tests presented in Fig. 3.12 prove that GPU-CASSERT scans the database much faster than the CPU-based implementation. Upon analyzing execution times for the first phase of the CASSERT algorithm (Fig. 3.12a, b) for both qualification thresholds, we can see that increasing the query protein's length causes the execution time for the algorithm to increase too. This is expected, since a longer query protein chain implies a longer alignment time for every pair of compared proteins. Small fluctuations that are visible for short chains when using the GPU-based implementation and $Q_T = 0.01$ (Fig. 3.12a, blue) are caused by variations in the number of secondary structures identified in the investigated proteins, which affect the alignment time. We can observe a similar (expected) dependency between the length of the query protein and the execution time while analyzing the measured execution times after both phases of the CASSERT algorithm for both qualification thresholds (Fig. 3.12c, d). However, since the number of proteins that qualify to the second phase varies and depends on the length and complexity of the query

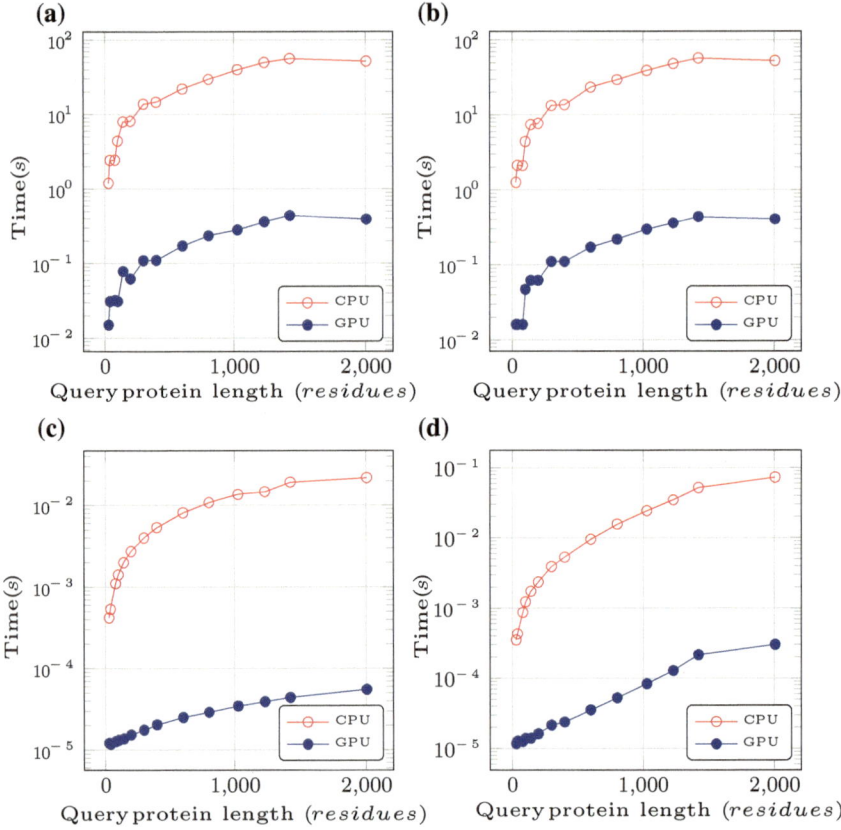

Fig. 3.12 Total execution time for the first phase (**a**, **b**) and average execution time of both phases per protein that qualified for the second phase (**c**, **d**) for qualification thresholds of 0.01 (**a**, **c**) and 0.2 (**b**, **d**) as a function of the length of the query protein structure Q. Time is plotted on a log_{10} scale. Comparison of two implementations of the CASSERT algorithm: CPU-based (*red*) and GPU-based (CUDA, *blue*). Results for 14 selected query protein structures between 29 and 2005 amino acids long. Searches were performed against the DALI database, containing 105,580 structures. **a** Execution time for the first phase for $Q_T = 0.01$. **b** Execution time for the first phase for $Q_T = 0.2$. **c** Average execution time per protein qualified for the second phase for $Q_T = 0.01$. **d** Average execution time per protein qualified for the second phase for $Q_T = 0.2$

structure, average execution times per qualified protein are shown in Fig. 3.12c, d. When executing the CASSERT for various query proteins, it can be noticed that, in some cases, more database protein structures qualify for the second phase for shorter query protein structures rather than longer (between 1000 and 2000 residues) query protein structures.

The execution time measurements that have been obtained during the performance tests allowed to calculate acceleration ratios for GPU-CASSERT with respect to CPU-CASSERT. Figure 3.13 shows how the acceleration ratio changes as a function

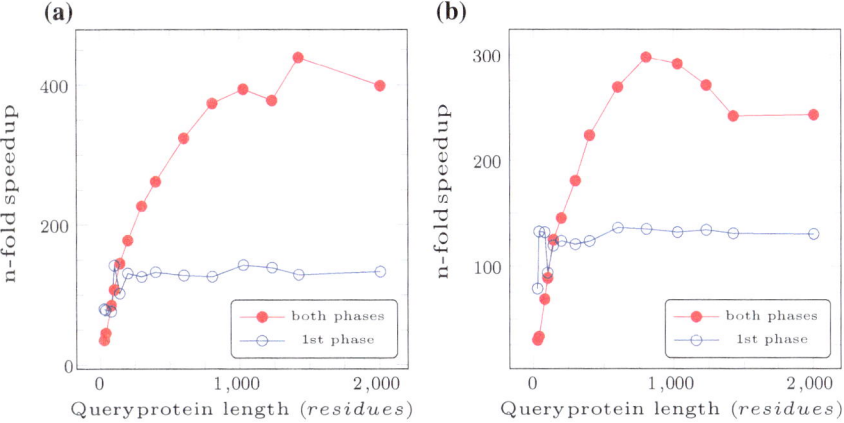

Fig. 3.13 Acceleration achieved by GPU-CASSERT with respect to CPU-CASSERT as a function of query protein length after the first phase (*blue*) and both alignment phases (*red*) with qualification thresholds 0.01 (**a**) and 0.2 (**b**). **a** Acceleration GPU versus CPU for $Q_T = 0.01$. **b** Acceleration GPU versus CPU for $Q_T = 0.2$

of query protein length for the first phase and both phases for $Q_T = 0.01$ and $Q_T = 0.2$.

We can see that the acceleration ratio for the first phase remains stable. In this phase, GPU-CASSERT is on average 120 times faster than CPU-CASSERT. However, for the whole alignment, i.e., after the first and second phases, the acceleration ratio greatly depends on the length of the query protein structure, its construction, and complexity. The whole alignment process when performed on the GPU is 30–300 times faster than the same process performed on the CPU.

Actually, for qualification thresholds $Q_T \geq 0.1$, it is possible to observe a kind of compensation effect. For longer query protein chains, which also have more complicated constructions in terms of secondary structure, the number of candidate structures from the database that qualified for the second phase decreases with the length of the query protein. This causes a situation in which fewer database proteins need to be aligned during the entire process. But, at the same time, the length of the query protein grows, causing the alignment time to increase. This growth is compensated for by the smaller number of database structures that need to be aligned.

Figure 3.14 shows the relationship between query protein length and the number of structures that qualified for the second phase when various values of the qualification threshold Q_T were applied. For example, for $Q_T = 0.01$ we can see that almost all of the database structures qualified for the second phase, regardless of query protein length. In this case, there is practically no filtering based on the secondary structures identified in the query protein. On the other hand, for $Q_T = 0.8$, we can notice that for query proteins over 150 residues in length, only single database structures are eligible for further processing.

In many situations, such a high value of the qualification threshold will filter out too many molecules. However, this depends on the situation for which the entire

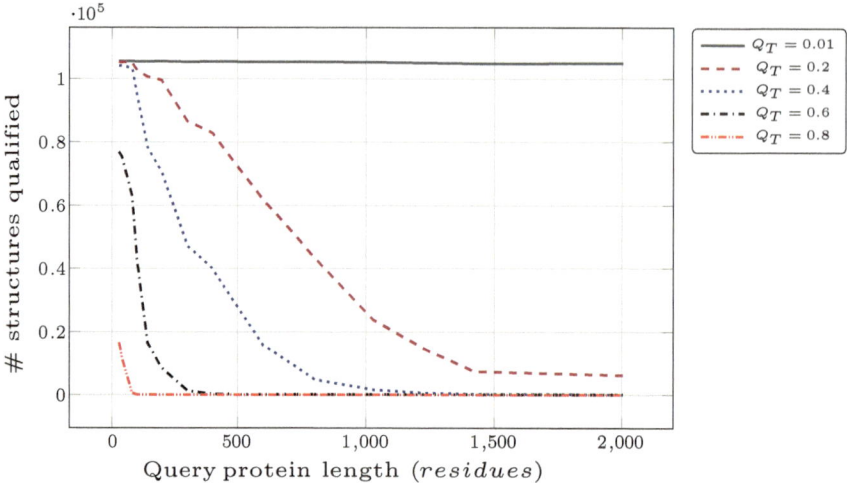

Fig. 3.14 Number of structures from the database that qualified for the second phase as a function of query protein length for various values of the qualification threshold

process of similarity searching is carried out. For example, in homology modeling, we may want to find referential protein structures that are very similar to the given query protein structure. For functional annotation and while searching for homologous structures, $Q_T = 0.2$ could be a reasonable threshold, since it filters out many candidate molecules and, even for very long query proteins, it allows several thousands of structures at least to pass through to the second phase.

We should also remember that the first alignment phase can be turned off completely by specifying $Q_T = 0.0$. Then, all of the database molecules pass through to the second phase, which prolongs the similarity searching process.

3.5 Discussion

The results of the efficiency tests have confirmed our expectations. Using a graphics card with a CUDA compute capability is one of the most efficient approaches to use when performing protein structure similarity searching. Upon comparing execution times, we can see that the GPU-based implementation is several dozen to several hundred times faster (an average of 180 times faster for $Q_T = 0.2$) than the CPU-based implementation. This is very important, since the number of protein structures in macromolecular databases, such as the Protein Data Bank, is growing very quickly, and the dynamics of this growth is also increasing. The use of GPU-based implementations is particularly convenient for such processes because GPU devices are reasonably inexpensive compared to, say, big computer clusters. Presented experiments were performed on a middle-class GPU device, which was set up on a small

PC workstation with two processors. For this reason, GPU devices can be usefully applied in the implementation of many algorithms in the field of bioinformatics.

The novelty of CPU-CASSERT lies mainly in the fast preselection phase based on secondary structures (the *low-resolution alignment* phase), which precedes the phase of detailed alignment (the *high-resolution alignment* phase). This allows the number of structures that will be processed in the second, costly phase to be limited, which, in turn, significantly accelerates the method itself. A comparison of CPU-CASSERT with the popular DALI and FATCAT algorithms is presented in [33].

GPU-CASSERT provides additional acceleration over its CPU-based version by executing the computational procedure in parallel threads on many cores of the GPU device. The resulting increase in speed is even greater than those achieved with the methods mentioned in the Sect. 3.1.3. *SA Tableau Search* provides a 33-fold increase in speed when using a GTX 285 graphics card and a 24-fold increase when using a C1060 GPU device rather than the CPU implementation. However, the optimization procedure is based on simulated annealing, which is run in parallel CUDA threads. Individual thread blocks perform the optimization procedure for different candidate protein structures from a database. Protein structures are represented as tableaux containing the orientations of secondary structure elements and distance matrices. However, one of the problems with this algorithm is encountered when comparing big protein structures that generate big tableaux and distance matrices, as they cannot be stored inside the constant and shared memory during computations. This makes it necessary to use a slower version of the GPU kernel which exploits the global memory rather than the faster constant and shared memory. GPU-CASSERT avoids this problem by using a different representation of protein structures: linear sequences of structural descriptors (where secondary structure elements are also included) are employed rather than two-dimensional representative structures.

In terms of representation of protein structures and the implementation of the method, GPU-CASSERT is closer to *pssAlign* [37], which shows up to a 35-fold increase in speed with the NVIDIA Tesla C2050 GPU over its CPU-based implementation. Both algorithms consists of two alignment phases. The *fragment-level alignment* phase of *pssAlign* uses an index-based matched fragment set (MFS) in order to find so-called seeds between the target protein and each database protein. These seeds, which are represented by the locations of the C_α atoms, are used to generate initial alignments which are then refined in the *residue-level alignment* phase. Just like GPU-CASSERT, both phases utilize dynamic programming. However, in GPU-CASSERT, the *low-resolution alignment* is treated as a preselection phase for detailed alignment. In contrast to *pssAlign*, both phases are executed independently in GPU-CASSERT. GPU-CASSERT does not store alignment paths after the first phase of the algorithm, which was done in the original CASSERT published in [33]. Consequently, it also does not perform backtracking in the kernel of the first phase, since GPU-CASSERT only needs the Score measure to calculate the qualification threshold Q_T for the next phase. The Score is calculated in a linear space, which also influences the effectiveness. Backtracking is also not performed in the GPU after the *high-resolution alignment* phase. It is executed on the host instead, and only for the highest-scoring database molecules that are returned for the user to visualize. This allows computational time to be saved.

Additional savings can be achieved when working with small query structures. After filtering candidate database proteins based on the qualification threshold, the program creates new, smaller data packages that are needed in the second phase. This usually takes some time. For this reason, for shorter query proteins (less than 100 amino acids in length), it is reasonable to omit the first phase by setting the qualification threshold to 0.0. The probability that such a small protein structure (after it has been reduced to a chain of SE regions) will be similar to many of the database proteins is very high. This means that all or almost all of the proteins qualify for the next phase (this is visible in Fig. 3.14), which makes the first preselection phase almost useless.

GPU-CASSERT also provides additional unique features. Following research into GPU-based sequence alignments [24, 25, 28, 39], the data are arranged in an appropriate manner before sending them to the global memory of the GPU device. Chains of structural descriptors representing protein structures are stored in a pre-pared memory array that guarantees coalesced access to the global memory in a single transaction. Structural descriptors are not transferred to the global memory of the GPU device directly from a database, but they are stored in binary files, which enables faster transfer, and they are sorted by their lengths in order to reduce thread idle time once they are processed. Moreover, secondary structure descriptors of query protein structures (in both phases) and residue types (in the second phase) are encoded as query profiles—appropriate matrices of all possible scores. During the computations performed on the GPU device, the query profile and substitution matrix (needed in the second phase) are located in the texture memory. The texture memory is cached on the chip of the graphics card and provides a higher effective bandwidth, reducing the number of requests made to off-chip global memory. Streaming is also applied in GPU-CASSERT in order to alternate kernel launches and memory copies, result-ing in further acceleration. Finally, kernel codes are optimized to avoid introducing branching via conditional statements.

3.6 Summary

Protein 3D structure similarity searching still needs efficient methods and new imple-mentations in order to generate results in a reasonable time. This has been prevalent taking into account exponentially growing numbers of protein structures in macro-molecular repositories. It seems that at the current stage of development of computer science, GPU devices provide an excellent alternative to very expensive computer infrastructures, as they allow large increases in speed over CPU-based implemen-tations for the same computational methods. Moreover, taking into account that the number of processing cores and the amount of memory in modern GPU devices are constantly growing, the computational capabilities of GPU devices are also growing at the same time. Although, implementing computational methods requires some additional effort by the user, including the need to get familiar with the completely new CUDA architecture and programming model, and to refactor the code of existing procedures into GPU kernels, in return we can achieve much faster processing. This is very important because, for many processes such as 3D protein structure similarity searching, reducing computational complexity is a very difficult, if not impossible, task. GPU-based implementations like that presented in the chapter do not reduce the complexity, but they can speed up the process by implementing massive parallelization, thus reducing the overall time required for process execution.

For the latest source codes of the GPU-CASSERT, please visit the project web site: http://zti.polsl.pl/dmrozek/science/gpucassert/cassert.htm

For further reading on GPU-based implementations of other algorithms for bioinformatics I would like to recommend the book entitled *Bioinformatics: High Performance Parallel Computer Architectures* by Bertil Schmidt [45]. In the next chapter, we will see how searching for 3D protein structure similarities against huge macromolecular repositories can be accelerated by using Cloud computing.

References

1. Bellman, R.: On the theory of dynamic programming. Proc. Natl. Acad. Sci. USA **38**(8), 716–719 (1952)
2. Berman, H., et al.: The Protein Data Bank. Nucleic Acids Res. **28**, 235–242 (2000)
3. Brown, N.P., Orengo, C.A., Taylor, W.R.: A protein structure comparison methodology. Comput. Chem. **20**, 359–380 (1996)
4. Brożek, M.: Protein structure similarity searching with the use of CUDA. MSc thesis, supervised by Mrozek D., Silesian University of Technology, Gliwice, Poland (2012)
5. Buckner, J., Wilson, J., Seligman, M., Athey, B., Watson, S., Meng, F.: The gputools package enbales GPU computing in R. Bioinformatics **26**, 134–135 (2010)
6. Burkowski, F.: Structural Bioinformatics: An Algorithmic Approach, 1st edn. Chapman and Hall/CRC, Boca Raton (2008)
7. Carugo, O.: Recent progress in measuring structural similarity between proteins. Curr. Protein Pept. Sci. **8**(3), 219–41 (2007)
8. Carugo, O., Pongor, S.: Recent progress in protein 3D structure comparison. Curr. Protein Pept. Sci. **3**(4), 441–449 (2002)
9. Can, T., Wang, Y.: CTSS: A robust and efficient method for protein structure alignment based on local geometrical and biological features. In: Proceedings of the 2003 IEEE Bioinformatics Conference (CSB 2003), pp. 169–179 (2003)
10. Coutsias, E.A., Seok, C., Dill, K.A.: Using quaternions to calculate RMSD. J. Comput. Chem. **25**(15), 1849–1857 (2004)
11. Daniluk, P., Lesyng, B.: A novel method to compare protein structures using local descriptors. BMC Bioinform. **12**, 344 (2011)
12. Friedrichs, M.S., Eastman, P., Vaidynathan, V., Houston, M., Legrand, S., Beberg, A.L., Ensign, D.L., Bruns, C.M., Pande, V.S.: Accelerating molecular dynamic simulation on graphics processing units. J. Comput. Chem. **30**(6), 864–872 (2009)
13. Gibrat, J., Madej, T., Bryant, S.: Surprising similarities in structure comparison. Curr. Opin. Struct. Biol. **6**(3), 377–385 (1996)
14. Godzik, A.: The structural alignment between two proteins: is there a unique answer? Protein Sci. **5**(7), 1325–1338 (1996)

15. Gu, J., Bourne, P.E.: Structural Bioinformatics (Methods of Biochemical Analysis), 2nd edn. Wiley-Blackwell, Hoboken, NJ (2009)
16. Henikoff, S., Henikoff, J.G.: Amino acid substitution matrices from protein blocks. Proc. Natl. Acad. Sci. USA **89**(22), 10915–10919 (1992)
17. Holm, L., Sander, C.: Protein structure comparison by alignment of distance matrices. J. Mol. Biol. **233**(1), 123–138 (1993)
18. Holm, L., Kaariainen, S., Rosenstrom, P., Schenkel, A.: Searching protein structure databases with DaliLite v. 3. Bioinformatics **24**, 2780–2781 (2008)
19. Horn, B.K.P.: Closed-form solution of absolute orientation using unit quaternions. J. Opt. Soc. Am. A **4**(4), 629–642 (1987)
20. Jamroz, M., Kolinski, A.: ClusCo: clustering and comparison of protein models. BMC Bioinform. **14**, 62 (2013)
21. Kabsch, W.: A solution for the best rotation to relate two sets of vectors. Acta Cryst. A **32**(5), 922–923 (1976)
22. Kabsch, W.: A discussion of the solution for the best rotation to relate two sets of vectors. Acta Cryst. **A34**, 827–828 (1978)
23. Lesk, A.M.: Introduction to Protein Science: Architecture, Function, and Genomics, 2nd edn. Oxford University Press, USA (2010)
24. Liu, Y., Maskell, D., Schmidt, B.: CUDASW++: optimizing Smith-Waterman sequence database searches for CUDA-enabled graphics processing units. BMC Res. Notes **2**, 73 (2009)
25. Liu, Y., Maskell, D., Schmidt, B.: CUDASW++2.0: enhanced Smith-Waterman protein database search on CUDA-enabled GPUs based on SIMT and virtualized SIMD abstractions. BMC Res. Notes **3**, 93 (2010)
26. Liu, Y., Wirawan, A., Schmidt, B.: CUDASW++ 3.0: accelerating Smith-Waterman protein database search by coupling CPU and GPU SIMD instructions. BMC Bioinform. **14**, 117 (2013)
27. Małysiak-Mrozek, B., Momot, A., Mrozek, D., Hera, Ł., Kozielski, S., Momot, M.: Scalable system for protein structure similarity searching. Lect. Notes Comput. Sci. **6923**, 271–280 (2011)
28. Manavski, S.A., Valle, G.: CUDA compatible GPU cards as efficient hardware accelerators for Smith-Waterman sequence alignment. BMC Bioinform. **9**, 1–9 (2008)
29. Minami, S., Sawada, K., Chikenji, G.: MICAN : a protein structure alignment algorithm that can handle multiple-chains, inverse alignments, Ca only models, alternative alignments, and Non-sequential alignments. BMC Bioinform. **14**, 24 (2013)
30. Momot, A., Małysiak-Mrozek, B., Kozielski, S., Mrozek, D., Hera, Ł., Górczyńska-Kosiorz, S., Momot, M.: Improving performance of protein structure similarity searching by distributing computations in hierarchical multi-agent system. Lect Notes Artif Int **6421**, 320–329 (2010)
31. Mosca, R., Brannetti, B., Schneider, T.R.: Alignment of protein structures in the presence of domain insertions. BMC Bioinform. **9**, 352 (2008)
32. Mrozek, D., Małysiak-Mrozek, B.: An improved method for protein similarity searching by alignment of fuzzy energy signatures. Int. J. Comput. Intell. Syst. **4**(1), 75–88 (2011)
33. Mrozek, D., Małysiak-Mrozek, B.: CASSERT: A two-phase alignment algorithm for matching 3D structures of proteins. In: Kwiecień A., Gaj P., Stera P. (eds.) CN 2013, CCIS, vol. 370, pp. 334–343 (2013)
34. Murzin, A.G., Brenner, S.E., Hubbard, T., Chothia, C.: SCOP: A structural classification of proteins database for the investigation of sequences and structures. J. Mol. Biol. **247**, 536–540 (1995)
35. Nvidia, CUDA C Programming Guide (Accessed on Aug 1, 2013) http://docs.nvidia.com/cuda/cuda-c-programming-guide/index.html
36. Ortiz, A.R., Strauss, C.E., Olmea, O.: MAMMOTH (matching molecular models obtained from theory): an automated method for model comparison. Protein Sci. **11**(11), 2606–2621 (2002)
37. Pang, B., Zhao, N., Becchi, M., Korkin, D., Shyu, C.-R.: Accelerating large-scale protein structure alignments with graphics processing units. BMC Res. Notes **5**, 116 (2012)

38. Pascual-Garca, A., Abia, D., Ortiz, A.R., Bastolla, U.: Cross-over between discrete and continuous protein structure space: insights into automatic classification and networks of protein structures. PLoS Comput. Biol. **5**(3), e1000331 (2009)
39. Pawłowski, R., Małysiak-Mrozek, B., Kozielski, S., Mrozek, D.: Fast and accurate similarity searching of biopolymer sequences with GPU and CUDA. Algorithms and Architectures for Parallel Processing, Lect Notes Comput Sci. **7016**, 230–243 (2011)
40. Roberts, E., Stone, J.E., Sepúlveda, L., Hwu W.M.W., Luthey-Schulten, Z.: Long time-scale simulations of in vivo diffusion using GPU hardware. In: IPDPS 09 Proceedings of the 2009 IEEE International Symposium on Parallel and Distributed Processing, pp. 1–8 (2009)
41. Rognes, T., Seeberg, E.: Six-fold speed-up of Smith-waterman sequence database searches using parallel processing on common microprocessors. Bioinformatics **16**, 699–706 (2000)
42. Sam, V., Tai, C.H., Garnier, J., Gibrat, J.F., Lee, B., Munson, P.J.: Towards an automatic classification of protein structural domains based on structural similarity. BMC Bioinform. **9**, 74 (2008)
43. Sanders, J., Kandrot, E.: CUDA by Example: An Introduction to General-Purpose GPU Programming, 1st edn. Addison-Wesley Professional, Pearson Education, Inc., Boston, MA (2010)
44. Schatz, M.C., Trapnell, C., Delcher, A.L., Varshney, A.: High-throughput sequence alignment using graphics processing units. BMC Bioinform. **8**, 474 (2007)
45. Schmidt, B.: Bioinformatics: High Performance Parallel Computer Architectures (Embedded Multi-Core Systems), 1st edn. CRC Press, Boca Raton, FL (2010)
46. Shapiro, J., Brutlag, D.: FoldMiner and LOCK2: protein structure comparison and motif discovery on the web. Nucleic Acids Res. **32**, 536–41 (2004)
47. Shindyalov, I., Bourne, P.: Protein structure alignment by incremental combinatorial extension (CE) of the optimal path. Protein Eng. **11**(9), 739–747 (1998)
48. Stanek, D., Mrozek, D., Małysiak-Mrozek, B.: MViewer: Visualization of protein molecular structures stored in the PDB, mmCIF and PDBML data formats. In: Kwiecień A., Gaj P., Stera P. (eds.) CN 2013, CCIS, vol. 370, pp. 323–333 (2013)
49. Stivala, A.D., Stuckey, P.J., Wirth, A.I.: Fast and accurate protein substructure searching with simulated annealing and GPUs. BMC Bioinform. **11**, 446 (2010)
50. Striemer, G.M., Akoglu, A.: Sequence alignment with GPU: performance and design challenges. In: IPDPS, IEEE International Symposium on Parallel and Distributed Processing, pp 1–10 (2009)
51. Suchard, M.A., Rambaut, A.: Many-core algorithms for statistical phylogenetics. Bioinformatics **25**(11), 1370–1376 (2009)
52. Ye, Y., Godzik, A.: Flexible structure alignment by chaining aligned fragment pairs allowing twists. Bioinformatics **19**(2), 246–255 (2003)
53. Yuan, C., Chen, H., Kihara, D.: Effective inter-residue contact definitions for accurate protein fold recognition. BMC Bioinform. **13**, 292 (2012)
54. Zemla, A.: LGA—a method for finding 3D similarities in protein structures. Nucleic Acids Res. **31**(13), 3370–3374 (2003)
55. Zhang, Y., Skolnick, J.: TM-align: a protein structure alignment algorithm based on the TM-score. Nucleic Acids Res. **33**(7), 2302–2309 (2005)
56. Zhu, J., Weng, Z.: FAST: a novel protein structure alignment algorithm. Proteins **58**, 618–627 (2005)

Chapter 4
Cloud Computing for 3D Protein Structure Alignment

> *Frankly, it is hard to predict what new capabilities the cloud may enable. The cloud has a trajectory that is hard to plot and a scope that reaches into so many aspects of our daily life that innovation can occur across a broad range.*

Barrie Sosinsky, 2011

Abstract Cloud computing provides huge amount of computational power that can be provisioned on a pay-as-you-go basis. In this chapter, we will see the cloud-based system for 3D protein structure alignment. The system was developed for the Microsoft Azure cloud and reached good, almost linearly proportional acceleration when scaled out onto many computational units. In this chapter, we will see that the alignment process can be successfully scaled out on cloud platforms.

Keywords Proteins · 3D protein structure · Tertiary structure · Similarity searching · Structure matching · Structure comparison · Structure alignment · Superposition · Cloud computing · Parallel computing · SaaS

4.1 Introduction

Popular methods for 3D protein structure similarity searching, like CE [14] and FATCAT [18], generate high quality structural alignments, but are still very time-consuming. As a consequence, the similarity searching against large repositories of structural data requires increased computational resources that are not available for everyone. Cloud computing provides huge amounts of computational power that can be provisioned on a pay-as-you-go basis. Cloud computing emerged as a result of requirements for the public availability of computing power, new technologies for data processing and the need of their global standardization, becoming a mechanism allowing to control the development of hardware and software resources by introducing the idea of virtualization. Cloud computing is a model that allows a

convenient, on-demand network access to a shared pool of configurable computing resources (e.g., networks, servers, storage, applications, and services) that can be
rapidly provisioned and released with minimal management effort or service provider
interaction [10]. In practice, cloud computing allows to run applications and services
on a distributed network using virtualized system and its resources, abstracting at
the same time from the implementation details of the system itself.

The use of cloud platforms can be particularly beneficial for companies and insti-
tutions that need to quickly gain access to a computer system which has a higher
than average computing power. In this case, the use of cloud computing services can
be more cost- effective and faster in implementation than using the owned resources
(servers and computing clusters) or buying new ones. For this reason, cloud computing is widely used in business and according to the Forbes [9] the market value
of such services will significantly increase in the coming years.

4.1.1 Cloud Computing in Bioinformatics and Life Sciences

The concept of cloud computing is also becoming increasingly popular in scientific
applications for which theoretically infinite resources of the cloud allow to solve the
computationally intensive problems. Also in the domain of bioinformatics, there are
many dedicated tools that are cloud-ready and several that have been created with
the aim of working in the cloud. Beneath we will see examples of some of the tools
and systems.

CloVR is a desktop application for automated sequence analysis that can utilize cloud computing resources. CloVR is implemented as a single portable virtual machine (VM) that provides several automated analysis pipelines for microbial
genomics, including 16S, whole genome and metagenome sequence analysis [1].
Cloud-based CloVR was developed for Amazon EC2 and automatically provisions
additional VM instances, if the computational process requires this. Hydra [8] is
an example of the cloud-ready tool that uses Hadoop and MapReduce in the identification of peptide sequences from spectra in the mass spectrometry. CloudBurst
[13] is a parallel read-mapping algorithm optimized for mapping next-generation
sequence data to the human genome and other reference genomes, for use in a
variety of biological analyses including SNP discovery, genotyping, and personal
genomics. CloudBurst uses Hadoop and MapReduce while parallelizing execution
using multiple compute nodes. Cloud-PLBS [3] implements the SMAP software for
3D ligand binding site comparison and similarity searching of a structural proteome
on the Hadoop framework using MapReduce paradigm. Cloud-PLBS parallelizes
the SMAP tool on a virtual cloud computing platform to handle the vast amount
of experimental data on protein-ligand binding site pairs [3]. Cloud BioLinux [7]
is a publicly accessible Virtual Machine (VM) that enables scientists to quickly
provision on-demand infrastructures for high-performance bioinformatics computing using cloud platforms. Users have instant access to a range of preconfigured
command line and graphical software applications, including a full-featured desktop

interface, documentation, and over 135 bioinformatics packages for applications including sequence alignment, clustering, assembly, display, editing, and phylogeny [7]. In the area of protein structure similarity searching, it is worth noting the work [4] by Che-Lun Hung and Yaw-Ling Lin and the PH2 system [5]. Authors of the first paper present the method for protein structure alignment and their own refinement algorithm that are implemented in Hadoop and are deployed on a virtualized computing environment. The PH2 system allows to store PDB files in a replicated way on the Hadoop Distributed File System and then allows to formulate SQL queries concerning 3D protein structures.

Cloud-based solutions for bioinformatics and life sciences usually relate to problems that require increased computational resources. As we know from the previous chapter, protein 3D structure similarity searching is one of the computationally complex and time-consuming processes. This motivates scientific efforts to develop scalable platforms that allow completing the task much faster. Cloud computing provides such a kind of scalable, high-performance computational platform.

4.1.2 Cloud Deployment and Service Models

Before we start describing a scalable, cloud-based solution for protein 3D structure similarity searching we need some background that will help us to position the solution in the cloud architecture. Developers of applications working in the cloud usually adopt one of the deployment models and develop the application to operate in one of the service models. In this chapter, we will take a look and explain types of deployment models and service models that can be adopted.

Deployment models decide where the infrastructure of the cloud will be located and managed, and who will use the cloud-based solution. We can distinguish here four widely accepted types [10]:

- Public cloud—the infrastructure of the cloud is available for public use or a large industry group and is owned by an organization selling cloud services;
- Private cloud—the cloud infrastructure is for the exclusive use of a single organization comprising multiple consumers (e.g., the organizational units); it does not matter whether it is a cloud managed by the organization, and it is located in its office; key factors for establishing private clouds seem to be: legal constraints, security, reliability, and lower costs for large organizations, and dedicated solutions;
- Community cloud—the cloud infrastructure is made available for the exclusive use of the consumer community from organizations that share common goals or are subjected to common legal restrictions;
- Hybrid cloud—the cloud infrastructure is based on a combination of two or more types of the above cloud infrastructures; if needed, allows for the use of public cloud resources to provide potential increased demand for resources (*cloud bursting*).

Fig. 4.1 Cloud services
defining types of components
that will be delivered to the
consumer

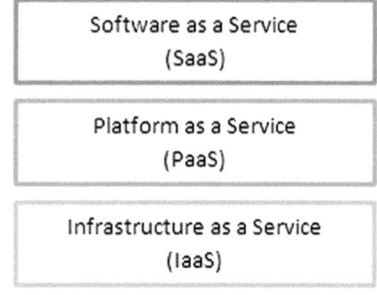

Service models define types of services that can be accessed on a cloud computing
platform. Among many others, three types of service are universally accepted. They
are usually presented in the form of stack as in Fig. 4.1.

The basis of the stack of services provided in clouds (Fig. 4.1) is the Infrastructure-
as-a-Service (IaaS) layer. IaaS provides basic computing resources in a virtualized
form, including: processing power, RAM, storage space, and appropriate bandwidth
for transferring data over the network, making it possible to deploy and run any
application. The IaaS service provider is responsible for the cloud infrastructure and
its management.

Platform-as-a-Service (PaaS) allows to create custom applications based on a
variety of services delivered by the cloud provider. As an addition to IaaS, the PaaS
provides operating systems, applications, development platforms, transactions, and
control structures. The cloud provider manages the infrastructure, operating systems,
and provided tools.

Software-as-a-Service (SaaS) provides services and applications with their user
interfaces that are available on an on-demand basis. The consumer is provided with an
application running in the cloud infrastructure. The consumer does not take care of the
infrastructure, operating systems, and other components underlying the application.
Its only responsibility is an appropriate interaction with the user interface and entering
appropriate data into the application. The user can also change the configuration of
the application and customize the user interface, if possible.

4.1.3 Microsoft Azure

Microsoft Azure is Microsoft's application platform for the public cloud. Microsoft
Azure provides services in the Platform-as-a-Service (PaaS) model and
Infrastructure-as-a-Service (IaaS) model, while capabilities of the Software-as-a-
Service (SaaS) model are offered by Microsoft Online Services.

In Fig. 4.2, we can see how a user benefits from the application deployed to the
Microsoft Azure cloud. The architecture consists of the following elements [17]:

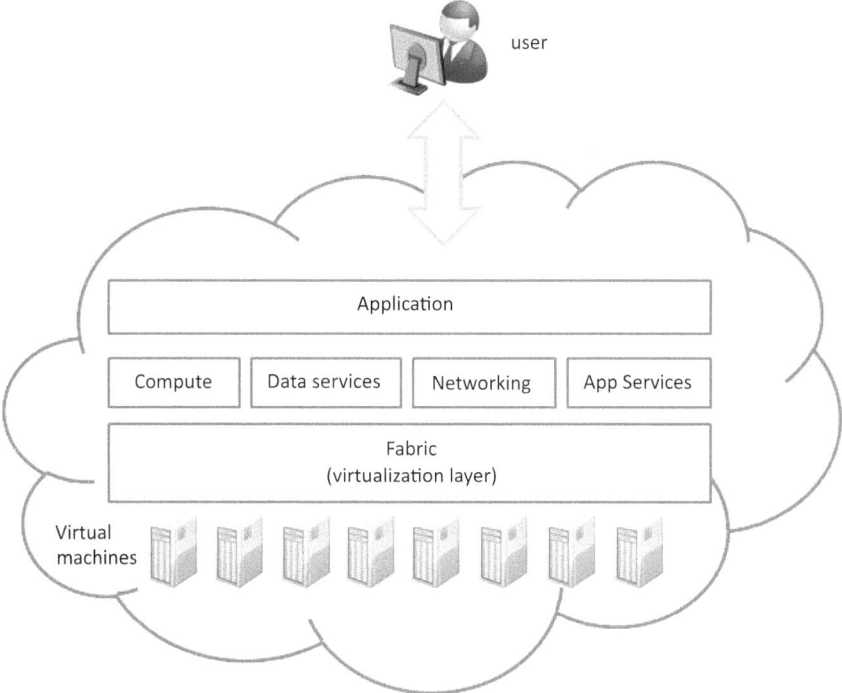

Fig. 4.2 Application deployed to Microsoft Azure which serves as a virtualized infrastructure, platform for developers, and gateway for hosting applications

- Application—the application that is made available to the community of users in the cloud, usually accessed by its web-based interface, Web service, or mobile interface.
- Compute—refers to compute capabilities of the Microsoft Azure platform providing separate services for particular needs:
  - Cloud Services—represent an application that is designed to run in the cloud service and XML configuration files that define how the cloud service should run; the application is defined in terms of component roles that implement the logic of the application; configuration files define the roles and resources for an application; the following roles can be used to implement the logic of the application:
    · Web role—is a virtual machine instance used for providing a Web-based front-end for the cloud service;
    · Worker role—is a virtual machine instance used for generalized development that performs background processing and scalable computations, accepts and responds to requests, and performs long running or intermittent tasks;
  - Virtual Machines—represent instances of virtual machines with preinstalled operating systems (Windows Server or Linux);

Table 4.1 Available sizes of Microsoft Azure virtual machines (VM) based on [16]

VM/server type	Number of CPU cores	CPU core speed (GHz)	Memory (GB)	Disk Space for Local Storage (GB)
Extra Small	Shared core	1.0	0.768	19
Small	1	1.6	1.75	224
Medium	2	1.6	3.5	489
Large	4	1.6	7	999
Extra Large	8	1.6	14	2039

- – Web Sites—represent websites that are created for business needs;
- – Mobile Services—represent highly functional mobile applications developed using Microsoft Azure;

- Data Services—provide the ability to store, modify, and report on data in Microsoft Azure; the following components of Data Services are provided:

  - – BLOBs—allow to store unstructured text or binary data (video, audio, and images);
  - – Tables—can store large amounts of unstructured nonrelational (NoSQL) data;
  - – SQL Database—formerly SQL Azure, allow to store large amounts of relational data;
  - – and others (SQL Data Sync, SQL Reporting, HDInsight).

- Networking—provide general connectivity and routing at the TCP/IP and DNS level;
- App Services—provide multiple services related to security, performance, work-flow management, and finally, messaging including Storage Queues and Service Bus providing efficient communication between application tiers running in Microsoft Azure.
- Fabric—entire compute, storage (e.g., hard drives), and network infrastructure, usually implemented as virtualized clusters; constitutes a resource pool that consists of one of more servers (called *scale units*).

Microsoft Azure platform allows to create five basic classes of virtual machines with different parameters and computational power (number of cores, CPU/core speed, amount of memory, efficiency of I/O channel). Table 4.1 shows a list of features for available computing units.

4.2 Cloud4Psi for 3D Protein Structure Alignment

One of the few systems in the world that utilizes cloud computing architecture to perform 3D protein structure similarity searching is Cloud4Psi (Fig. 4.3). The system can be scaled out (horizontal scaling) and scaled up (vertical scaling) in the Microsoft

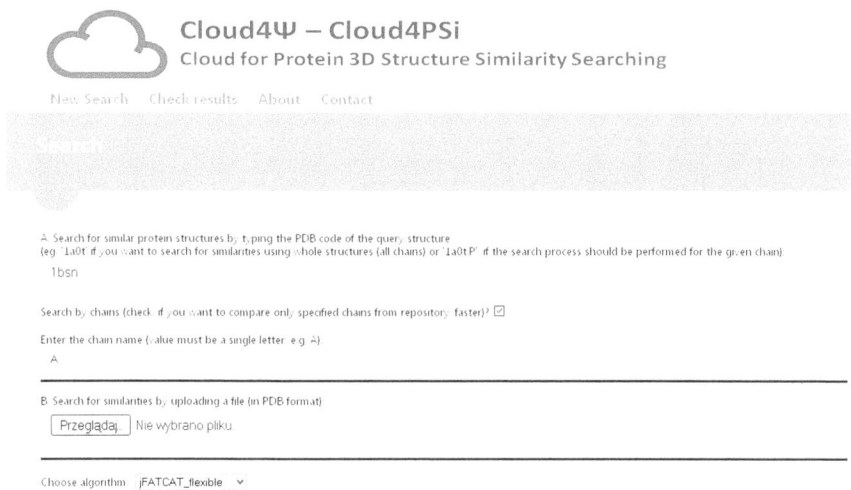

Fig. 4.3 Cloud4Psi web site. For the latest information on the Cloud4Psi, please visit the project home page: http://zti.polsl.pl/dmrozek/science/cloud4psi.htm

Azure public cloud. Scaling up allows for the expansion of computational resources like increasing the number of processor cores, adding more memory, or moving the workload to the computation unit possessing better performance parameters. Horizontal scaling or scaling out, is achieved by increasing the number of the same units and appropriate allocation of tasks among these units. Microsoft Azure allows to combine both types of scaling. It is worth noting that in case of Cloud4Psi, vertical scalability required designing and implementation of the application code in such a way that it utilizes many processing cores available after scaling the system up. For the horizontal scalability, the Cloud4Psi code had to be properly designed in order to allow the division of tasks between computation units and operate with restrictions on sharing resources (mainly memory).

It is worth mentioning that a part of the work was carried out in the cooperation with Kłapciński [6], my associate in this project, and is continued by me and my research group under the Microsoft Azure for Research Award program sponsored by Microsoft Research. The project is entitled *Cloud4Psi: Cloud Computing in the Service of 3D Protein Structure Similarity Searching*.

4.2.1 Use Case: Interaction with Cloud4Psi

Users may interact with the Cloud4Psi system on two levels. They may generate requests for the similarity searching or they may configure Cloud4Psi, e.g., when they want to scale up or scale out the system (Fig. 4.4). Execution of the similarity searching and displaying its results remains the basic scenario implemented by the

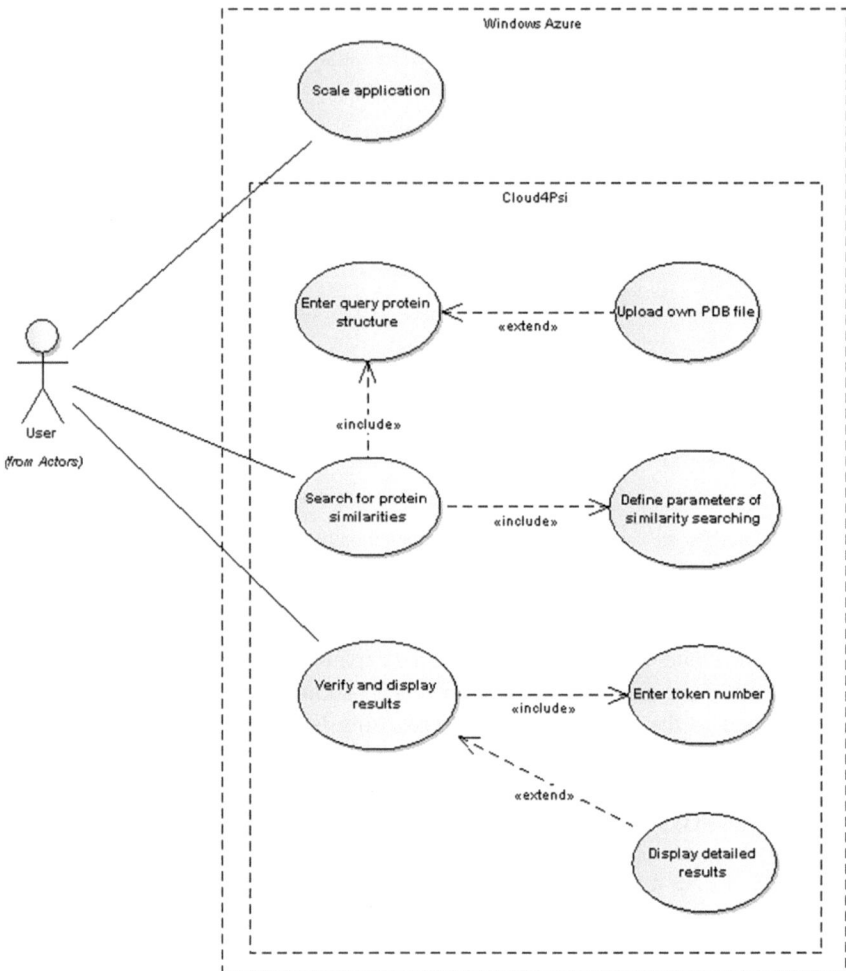

Fig. 4.4 Interaction of the user with the Cloud4Psi system. Typical use cases for similarity searching and scaling the system

Cloud4Psi. The process includes entering through a dedicated website a query protein structure, either by providing the PDB ID code with or without the chain identifier, or by uploading user's structure from a local hard drive to the Cloud4Psi storage system. Users may also choose one of the available algorithms for the similarity searching (jCE, jFATCAT rigid, jFATCAT flexible) and define parameters of the process, if they do not want to use default values.

A special token number is generated for the user for each search submission. The token number can be used in order to get partial or full results of the submitted search after some time. Users do not have to follow changes on the website, since the similarity searching can be a long running process. They can return to the website at any convenient moment and check, if the process has already been completed.

The configuration and scaling of the Cloud4Psi are mainly reserved for advanced users of the system and can be performed outside of the Cloud4Psi itself, e.g., in Microsoft Azure management console. Cloud4Psi can be scaled up by raising the capabilities of compute units (according to Tab. 4.1) or scaled out by adding more searching instances.

4.2.2 Architecture and Model of the Cloud4Psi

As we know from the Sect. 4.1.3, any application that runs in the Microsoft Azure cloud is composed of a set of roles performing some tasks. Breakdown of the roles depends on process that is implemented and delivered by the cloud-based application. Cloud4Psi consists of several types of roles and storage modules responsible for gathering and exchanging data between computing roles.

The set of roles $R_{C4\psi}$ working in the Cloud4Psi system is defined as follows:

$$R_{C4\psi} = \{r_W\} \cup \{r_M\} \cup R_S, \tag{4.1}$$

where r_W is the Web role responsible for the interaction with Cloud4Psi users, r_M is the Manager role distributing requests received from the Web role and preparing the workload, and R_S is the set of Searcher roles performing embarrassingly parallel structural alignments.

The Web role provides Graphical User Interface (GUI) through a friendly website and consists of additional logical layer for even handling. Through the Web role users can initiate the similarity search or receive the results of the process already initiated or completed. Logical layer is responsible for converting parameters received from the user to a format that can be transmitted to the Manager role through the Input queue (Fig. 4.5). The Web role has an access to the Storage Tables that provide results of the similarity searches for displaying purposes. It also has an access to the virtual hard disk storing PDB files, when the user decides to send its own PDB file to be compared by the Cloud4Psi.

Manager role is one of the Worker roles. It distributes requests received from the Web role, passes parameters, arranges the scope of the similarity searching, and manages associated computational load between Searcher roles. Manager is also responsible for the preparation of the read-only, virtual hard disk, which will be used by Searcher roles.

Searcher roles, which are also Worker roles, bear the computational load associated with the process of protein comparison and alignment. The set of Searcher roles R_S is defined as follows:

$$R_S = \{r_{Si}|i = 1, ..., n\} \tag{4.2}$$

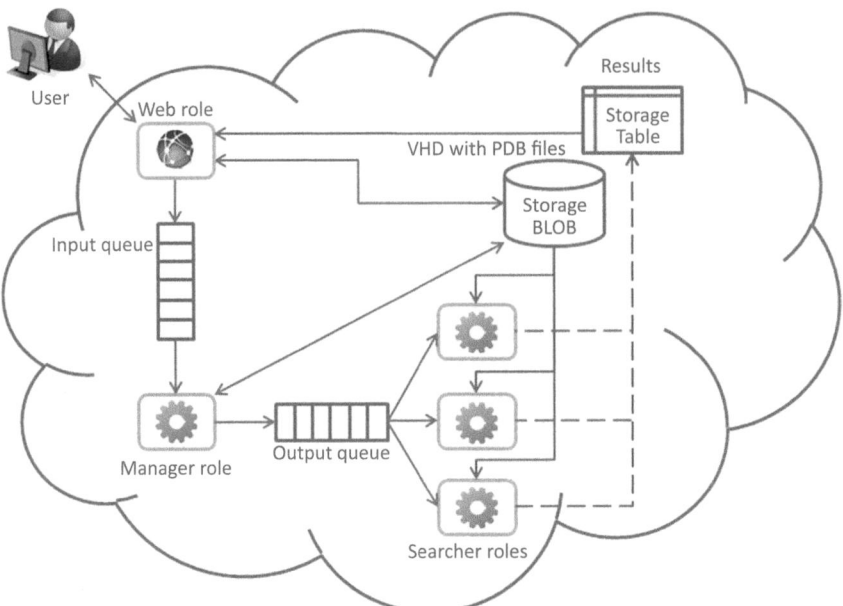

Fig. 4.5 Architecture of the Cloud4Psi—a Microsoft Azure cloud-based solution for protein structure similarity searching: Web role provides the front-end for users of the system, Manager role
mediates the distribution of the searching process, which is executed by Searcher roles. Tasks that
should be completed are transferred through Input and Output queues. Roles have an access to
various storage resources, including Storage BLOB and Storage Tables

where r_{Si} is a single Searcher role, and n is the number of Searchers working in the
system.

Each Searcher role r_{Si} performs batch comparisons of the given protein structure
with subsets of proteins from the repository P_{DB}. This role type can be scaled out
and scaled up during the similarity searching process (vertical and horizontal scaling
of computational system). When scaling out, users of the Cloud4Psi change the value
of n in formula 4.2. When scaling up, users change the size of the Searcher role:

$$size(r_{Si}) \in \{XS, S, M, L, XL\}, \tag{4.3}$$

where XS denotes extra small size, S—small, M—medium, L—large, XL—extra
large. Sizes of the Searcher roles are the same as those described in Table 4.1 and
are consistent with parameters of the provided computation units.

The following assumption regarding Searcher roles is valid in the Cloud4Psi:

$$\forall_{r_{Si},r_{Sj}\in R_S, i,j\in1,...,n,i\neq j} \quad size(r_{Si}) = size(r_{Sj}). \tag{4.4}$$

The number of instances n, on which the Searcher role can run, depends only on the user's choice and the range of services and resources that are provided by the Microsoft company as the owner of the Microsoft Azure cloud. Searcher roles receive from the Manager role messages with the information on the scope of the main tasks that should be performed by the particular Searcher, the name of the comparison algorithm that should be used, and a list of PDB files to compare from a virtual hard drive. The list of protein structures (PDB files) is called a *package*. Finally, Searcher roles are responsible for entering results to a table in the Storage Table service.

Packages that are sent to Searcher roles consist of lists of structures from the main repository P_{DB} of PDB files:

$$P_{DB} = \bigcup_{i=1}^{m} p_i, \tag{4.5}$$

where p_i is the i-th package of protein structures, and m is the number of packages that should be processed by all Searcher roles. Packages satisfy the following relationship:

$$\forall_{1 \le i, j \le m} \quad i \ne j \implies p_i \cap p_j = \emptyset. \tag{4.6}$$

The m depends on the repository size and the size of a package:

$$m = \left\lceil \frac{size(P_{DB})}{size(p_i)} \right\rceil. \tag{4.7}$$

The size of repository may include all protein structures that are available in the Protein Data Bank [2] or can be restricted just to chosen structures. The size of the package p_i must be chosen experimentally.

The whole computing architecture of the Cloud4Psi system for protein structure similarity searching is shown in Fig. 4.5. The system consists of three types of roles mentioned in the model and additional modules responsible for storing and exchanging data. These are the following:

- Table (Storage Table service), which stores the results of the similarity searching, time stamps of the key moments of the application run (needed when studying performance of the system), and technical parameters used globally by all roles.
- A pageable BLOB (Storage BLOBs service) that contains the virtual disk (VHD). The disk is mounted by the Web role in the *full* mode, if the user chooses to upload its own protein PDB file as a query structure, or in the *read-only* mode as the current image of the PDB repository for Searcher roles performing parallel, distributed similarity searches.
- Input queue, which collects similarity searching requests from the Web role and provides these requests to the Manager role, where they are distributed among the instances of the Searcher role.
- Output queue, which stores messages with the information on what part of the PDB repository should be processed by the Searcher role that receives particular message and comparison parameters.

4.2.3 Algorithms for Protein Structure Similarity Searching

Cloud4Psi allows searching for protein structure similarities by means of two algorithms—FATCAT [18] and CE [14]. Actually, it uses new, enhanced implementations of these algorithms, called jFATCAT and jCE, published in [12]. Both algorithms have a very well-established position among researchers and are publicly available through the Protein Data Bank website for those, who want to search for structural neighbors. Moreover, both algorithms are used for precalculated 3D-structure comparisons for the whole PDB that are updated on a weekly basis [11]. FATCAT and CE work on the basis of matching protein structures using Aligned Fragment Pairs (AFPs) representing parts of protein structures that fit to each other. However, FATCAT eliminates drawbacks of many existing methods that treat proteins as rigid bodies, not flexible structures. The research conducted by the authors of the FATCAT has shown that rigid representation causes that a lot of similarities, even very strong, are omitted. On the other hand, FATCAT allows to enter twists in protein structures while matching their fragments providing better alignments in a number of cases. One of the cases is shown in Fig. 4.6. It shows how two protein structures are aligned using the CE algorithm and the FATCAT algorithm.

In Fig. 4.6a we can see structures of proteins [PDB ID: 2SPC.A] and [PDB ID: 1AJ3.A] and their alignment generated by the CE algorithm, which treats structures as rigid bodies. Colors point out parts of the structures that were aligned. Both structures are highly homologous, which is also reflected in their sequence alignment. However, a different orientation of the rest of compared chains causes that these parts are not regarded as structurally similar. This applies not only to the CE algorithm, but also other algorithms treating proteins as rigid bodies. FATCAT is able to handle such deformations and various orientations by entering gaps and twists (rigid body movements). Appropriate penalty system is used in order to limit the number of these operations. In Fig. 4.6b we can see the structural alignment of the same two structures after entering gaps and twists. As a result, FATCAT finds new regions reflecting structural similarity.

4.2.4 Implementation of Similarity Searching in Azure Cloud

Now let us go deeper into the implementation details of the system. Cloud4Psi allows users to execute the similarity searching through a dedicated website. The website is provided by the Web role. A user inputs a query protein structure, either by PDB ID or by uploading user's protein structure from the local hard drive, and chooses one of the algorithms for similarity searching (jCE, jFATCAT rigid, or jFATCAT flexible). Additionally, the user can specify parameters of the chosen algorithm. When the user starts the search process, the Web role behaves according to the pseudocode of the Algorithm 1. It generates a token number for the search request (line 2). The user may return to the website after sometime with this token number, and check

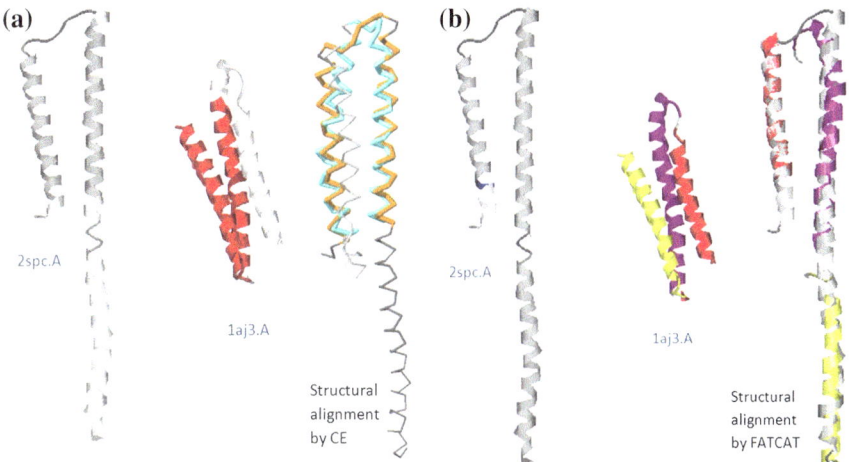

Fig. 4.6 Superposition of proteins [PDB ID: 2SPC.A] and [PDB ID: 1AJ3.A] after structural alignments generated by the CE algorithm (**a**) and the FATCAT algorithm (**b**). Parts of the structures that were structurally aligned are marked by using various colors. In the case of the FATCAT (**b**), the structure [PDB ID: 2SPC.A] is transformed by entering a twist, which gives better alignment

whether the search process has already finished. If the user chose to upload his own structure as the query protein, the Web role mounts the virtual hard drive in the *full* mode, uploads the structure to the hard drive, and encodes the locator to the structure in the search request message (lines 3-6). If the user chose to search similarities by providing PDB ID code of the query protein structure, the Web role encodes the code in the search request message (lines 7-8). Then, the Web role generates the search request message that is sent to the Input queue (line 10). The format of the input message is presented in Listing 4.1.

Algorithm 1 Web role: Search request processing algorithm

1: **for each** search request **do**
2:     Return token number
3:     **if** user uploads protein structure **then**
4:         Mount virtual hard drive (VHD) in the *full* mode
5:         Upload user's query protein structure
6:         Encode the locator of user's protein in the search request message
7:     **else**
8:         Encode PDB ID code of the query protein in the search request message
9:     **end if**
10:    Enqueue the search request in the Input queue
11: **end for**

```
 1  CloudQueueMessage  searchRequest  =  new  CloudQueueMessage(
 2                          guid.ToString() + "|"
 3                          + part_size + "|"
 4                          + struct_num + "|"
 5                          + pdb_id + "|"
 6                          + upload_name + "|"
 7                          + messageTime + "|"
 8                          + algorithm + "|"
 9                          + byChain.Checked.ToString());
10                      inputQueue.AddMessage(searchRequest);
```

Listing 4.1 Format of the input message for search request (based on [6])

Input and Output queues in the Cloud4Psi use text messages. Particular components of the text messages are separated using the | symbol. The input message consists of eight component parts:

- a randomly generated token number (*guid*), which is returned to the user,
- the number of proteins in the package that should be compared by each instance of the Searcher role in a single iteration (*part_size*),
- the number of proteins from the repository that will be used in the whole comparison process (*struct_num*, used for performance tests),
- the *pdb_id* identifier of the user's query protein to quickly locate it in the repository,
- the locator of the query protein structure that was uploaded from the user's computer (*upload_name*),
- a marker defining the time of dispatch of the search request (*messageTime*, used for time statistics),
- encoded name of the algorithm used for similarity searching (*algorithm*), and finally,
- the information whether the comparison is performed by using whole protein structures or just between selected chains (*ByChain*).

Some of the components (e.g., *guid*) are used later while accessing Storage Tables service to identify the specific outcome of the search job.

Manager Worker realizes the pseudocode of Algorithm 2. The role listens if there are any search requests in the Input queue (line 2). Incoming requests are immediately captured by the Manager Worker role, which divides the whole range of repository molecules into packages (lines 3-5). Packages contain a small number of protein structures that should be compared with user's query protein by a single Searcher role. Descriptors of successive packages are sent by Manager worker role as messages to the Output queue where they wait for being processed (lines 6–9). The format of the message that is sent to the Output queue is shown in Listing 4.2.

Algorithm 2 Manager role: Search request processing and package generation algorithm

---

1: **while** true **do**
2:     Check messages in the Input queue
3:     **if** exists a message **then**
4:         Retrieve the message and extract parameters
5:         Divide repository P_{DB} into smaller packages according to the defined package size
6:         **for each** package $p_i \subset P_{DB}$ **do**
7:             Encode package metadata and other parameters in the output message
8:             Enqueue the output message in the Output queue
9:         **end for**
10:    **end if**
11: **end while**

```
1  CloudQueueMessage packageDescriptor = new CloudQueueMessage(
2                            token + "|"
3                            + part_size + "|"
4                            + struct_num + "|"
5                            + pdb_id + "|"
6                            + upload_name + "|"
7                            + start_point + "|"
8                            + messageTime + "|"
9                            + snapshotUri + "|"
10                           + algorithm + "|"
11                           + byChain
12                           );
13                outputQueue.AddMessage(packageDescriptor);
```

Listing 4.2 Format of a message containing package descriptor that is sent to the Output queue (based on [6])

Sample message from the Output queue is presented below:

```
4aa284b0-d1aa-42bc-9f1d-4cfcf98b2641|10|100000|1bsn.A|True|20|
3/10/2014 8:53:55 AM|
http://prot.blob.core.windows.net/drives/pdb.vhd?snapshot=...|
1|True
```

Most of the information stored in the message comes from the search request message and is just forwarded to Searcher worker roles. Additionally, the package descriptor message consists of the URL address of the mounted read-only, virtual hard drive image containing P_{DB} repository with protein structures (*snapshotUri*). PDB identifiers of structures from the repository are placed in the file array of the size *struct_num*. Searcher worker roles process packages containing the number of structures determined by the *part_size* parameter starting from the *start_point*. The *start_point* defines an offset in the P_{DB} repository for each Searcher worker role (Fig. 4.7). In Fig. 4.7 each square can be interpreted as a part of the whole P_{DB} repository (a package), which description was sent to the Output queue and will be retrieved by a single instance of the Searcher worker role. For each package descriptor that is generated by the Manager worker role the *start_point* is incremented by the value of the package size (*part_size*).

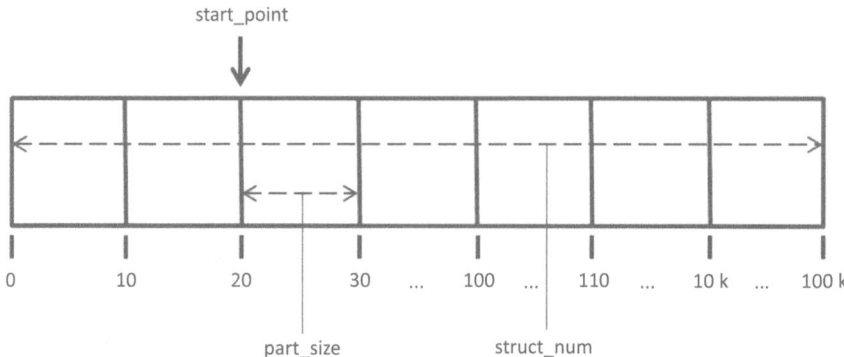

Fig. 4.7 Division of the P_{DB} repository based on the information in a sample message: repository of the size *struct_num = 100,000*, package size *part_size = 10*, *start_point = 20* for one of the Searcher roles that will retrieve the message from the Output queue (based on [6])

Each Searcher Worker role processes the package by comparing the query protein structure to structures which identifiers are contained in the package (Algorithm 3). Identifiers of the query protein structure and candidate structures are passed in the package descriptor message, as well as the name of comparison algorithm (mapped to integer) that should be used (line 4).

Algorithm 3 Searcher role: Package processing algorithm

1: **while** true **do**
2:     Check messages in the Output queue
3:     **if** exists a message **then**
4:         Retrieve the message and extract parameters
5:         Get query protein structure S^Q from virtual hard drive
6:         **for each** database structure $S^D \in p_i$ **do**
7:             Get the candidate database structure S^D from P_{DB} repository on virtual hard drive
8:             Compare structures S^Q, S^D with the use of selected algorithm
9:             Collect comparison results in a dedicated array
10:        **end for**
11:        Save collected results in the Storage Table
12:    **end if**
13: **end while**

If the Searcher worker role operates on the compute unit possessing many cores, all cores of the compute unit are used (the task is parallelized inside the Searcher role). Candidate protein structures described by a package descriptor message are taken from the virtual, read-only hard drive located in the Storage Drive service of the Azure cloud (line 6-7). After comparing all structures in the package (lines 8-9), outcomes of the comparison, i.e., PDB identifiers of structures and similarity measures, are sent to the table of results available through the Azure Storage Tables service (line 11).

Fig. 4.8 Retrieving similarity searching results from the Cloud4Psi

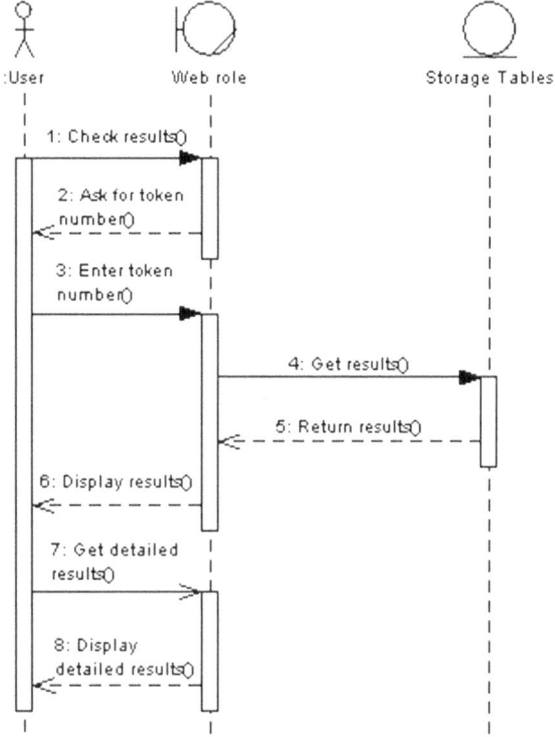

Instances of the Searcher worker role work in a loop. After processing a package the role returns to listening and capturing messages from the Output queue (lines 1-4). Successive packages are processed until there are no more messages in the queue.

Web role allows users to check results of the similarity searching through appropriate web form. The Web role asks the user to provide the token number that was generated during the execution of the process (Fig. 4.8). Results that are assigned to the given token number are then retrieved from the Storage Tables service and are displayed to the user.

These results include identifiers of protein structures (sorted by a chosen similarity measure) and similarity measures specific for the similarity algorithm, e.g., Z-score, RMSD, alignment length, P-value, TM-score, and others. The user may also display detailed structural alignment report for a pair of query protein structure and selected database structure returned by the Cloud4Psi. Sample detailed structural alignment report is shown in Fig. 4.9.

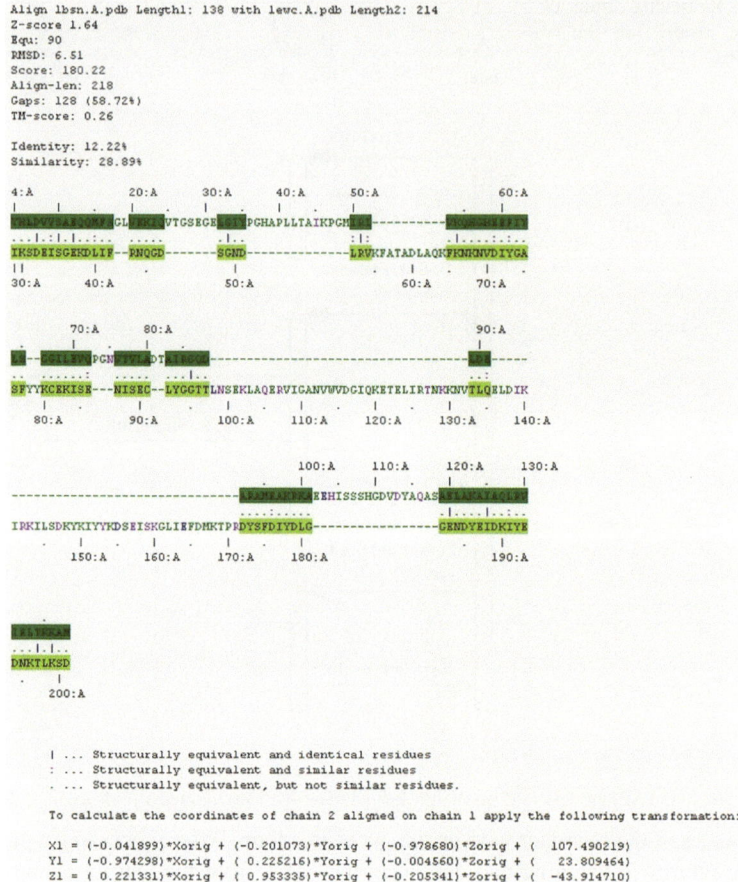

Fig. 4.9 Detailed report showing structural alignment of sample protein structures [PDB ID: 1BSN.A] and [PDB ID: 1EWC.A]. Parts of chains marked using dark green and light green colors reflect regions of structures that correspond to each other. A vertical line between residues reflects structural equivalence and identical residues, a colon means structural equivalence and similar residues, and a dot means structural equivalence, but not similar residues

4.3 Efficiency of the Cloud4Psi

In order to assess the performance of the presented architecture, the Cloud4Psi has undergone a series of tests. During these tests the system was mainly scaled horizontally. In particular, we have examined the efficiency of the similarity searches depending on the number of instances of the Searcher role. During all tests, the Web role and the Manager role were running on computational units of the *Small* size, and sizes of the computational units for the Searcher role were variable in different experiments. However, horizontal scaling with the use of Searchers of the *Small* size

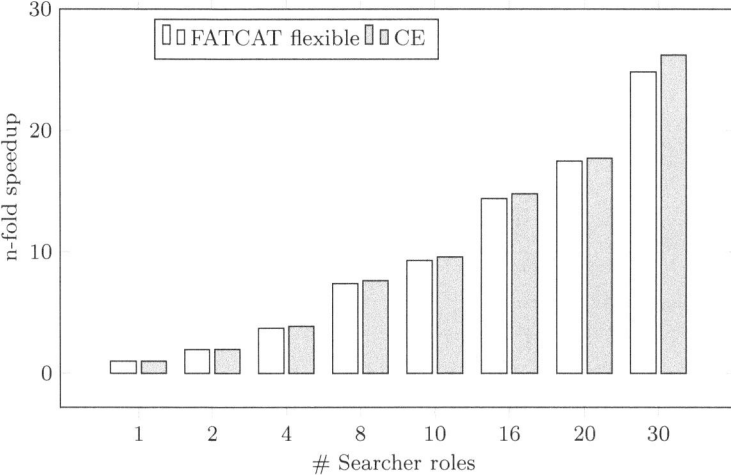

Fig. 4.10 Acceleration of the similarity searching as a function of the number of Searcher roles for CE and FATCAT algorithms

turned out to be more elastic and easier in deployment. Scaling out can be performed from the Microsoft Azure management console without republishing the system from the development environment. For this reason, we will see results for this type of scaling. Tests were conducted for various protein structures. The results were averaged. Similarity searching was carried out with the use of both algorithms, jCE and jFATCAT (flexible), against the P_{DB} repository containing one thousand different structures from the Protein Data Bank. Package size was experimentally set to 10 protein structures ($size(p_i) = 10$), which guarantees reasonable load balancing and flexibility in assigning packages to Searcher roles.

In Fig. 4.10, we can observe acceleration of the similarity searching as a function of the number of Searcher roles for both algorithms. We can notice that employing two instances of the Searcher role speeds up the process almost twice.

Adding more Searcher roles further accelerates the process. However, the dynamics of the acceleration slowly decreases. Finally, by increasing the number of Searcher roles from 1 to 30, we gained the average n-fold speedup at the level of 26.24 for the jCE algorithm, and 24.84 for the jFATCAT algorithm.

Although, based on the execution time measurements we have noticed that the jCE algorithm is 25–40 % faster than jFATCAT (both variants of the jFATCAT have similar execution times), we can see that the acceleration ratios are similar for both algorithms. This indicates that the scalability of the Cloud4Psi does not depend on the algorithm, but it depends on the assumed architecture and system components. In other words, there are some other factors influencing the dynamics of the acceleration. Among these factors we can distinguish the communication between roles through the queueing system, necessity of retrieving query protein structure from repository located in the virtual hard drive and the need to store partial results of the similarity searching for each processed package in the Storage Table.

4.4 Discussion

Cloud4Psi represents a novel architectural approach in building cloud-based systems for protein structure similarity searching and bioinformatics by implementing dedicated role-based and queue-based model. Most solutions developed so far are mainly based on preconfigured virtual machines. Their images can be setup in a cloud, if a user wants to scale out the computational process. However, these solutions does not provide full features of the SaaS layer.

Cloud4Psi is a fully SaaS solution. It requires the user gets familiar just with the web-based Graphical User Interface (GUI), and everything else is hidden under the GUI. From the viewpoint of the maintenance of the system, the role-based model applied in the Cloud4Psi provides higher portability (inside the same cloud provider) and significantly higher flexibility in deployment of the Cloud4Psi on various operating systems.

An interesting alternative for such a processing problems provide systems that are built based on the Hadoop platform, like the one developed by Che-Lun Hung and Yaw-Ling Lin and reported in [4]. This system, however, represents a different approach to the parallel implementation of the similarity searching process, which is based on the MapReduce paradigm. Cloud4Psi uses its own dedicated scheduling architecture with various types of roles and queues. Using queues has several advantages. Since queues provide asynchronous messaging model, users of the Cloud4Psi need not be online at the same time. Queues reliably store requests as messages until the Cloud4Psi is ready to process them. Cloud4Psi can be adjusted and scaled out according to the current needs. As the dept of the request queue grows, more computing resources can be provisioned. Therefore, such an approach allows to save money taking into account the amount of infrastructure required to service the application load. Finally, queue-based approach allows load balancing—as the number of requests in a queue increases, more Searcher roles can be added to read from the queue.

4.5 Summary

Cloud allows for immediate and temporary lease of publically available computing infrastructure, according to current needs and without having to purchase an expensive equipment. This gives the possibility to quickly enter the market, make the product available to users, and scale it on demand with the increasing requirements.

Cloud4Psi benefits from the idea of cloud computing by utilizing computation units to scale the process of 3D protein structure similarity searching—the process that is time-consuming and very important from the perspective of structural bioinformatics, comparative genomics, and computational biology. In this chapter, we could observe that scaling the process in the cloud improves the efficiency of the search process without reducing the computational complexity of used alignment methods.

For further reading on cloud computing I would like to recommend the book entitled *Cloud Computing Bible* by Sosinsky [15]. For readers that are interested in various applications of Hadoop framework and MapReduce programming model in bioinformatics I recommend the fresh paper of Zou et al. [19]. In Chap. 5, we will briefly summarize the technologies used to build the solutions presented throughout the book.

References

1. Angiuoli, S.V., Matalka, M., Gussman, A., Galens, K., et al.: CloVR: A virtual machine for automated and portable sequence analysis from the desktop using cloud computing. BMC Bioinformatics **12**, 356 (2011)
2. Berman, H., et al.: The Protein Data Bank. Nucleic Acids Res. **28**, 235–242 (2000)
3. Hung, C-L., Hua, G-J.: Cloud Computing for Protein-Ligand Binding Site Comparison. Biomed Res Int. 170356 (2013)
4. Hung, C.-L., Lin, Y.-L.: Implementation of a parallel protein structure alignment service on cloud. Int. J. Genomics **439681**, 1–8 (2013)
5. Hazelhurst, S.: PH2: an hadoop-based framework for mining structural properties from the PDB database. In: Proceedings of the 2010 Annual Research Conference of the South African Institute of Computer Scientists and Information Technologists, pp. 104–112 (2010)
6. Kłapciński, A.: Scaling the process of protein structure similarity searching in cloud computing. MSc thesis, supervised by Mrozek D., Institute of Informatics, Silesian University of Technology, Gliwice, Poland (2013)
7. Krampis, K., Booth, T., Chapman, B., Tiwari, B., et al.: Cloud BioLinux: pre-configured and on-demand bioinformatics computing for the genomics community. BMC Bioinform. **13**, 42 (2012)
8. Lewis, S., Csordas, A., Killcoyne, S., Hermjakob, H., et al.: Hydra: a scalable proteomic search engine which utilizes the Hadoop distributed computing framework. BMC Bioinform. **13**, 324 (2012)
9. McKendrick, J.: Cloud Computing Market Hot, But How Hot? Estimates are All Over the Map. Forbes, http://www.forbes.com/sites/joemckendrick/2012/02/13/cloud-computing-market-hot-but-how-hot-estimates-are-all-over-the-map/ (2012). Accessed 25 Nov 2013
10. Mell, P., Grance, T.: The NIST Definition of Cloud Computing. Special Publication 800–145. http://csrc.nist.gov/publications/nistpubs/800-145/SP800-145.pdf (2011). Accessed 25 Nov 2013
11. Prlić, A., Bliven, S., Rose, P.W., Bluhm, W.F., Bizon, C., Godzik, A., Bourne, P.E.: Precalculated protein structure alignments at the RCSB PDB website. Bioinformatics **26**, 2983–2985 (2010)
12. Prlić, A., Yates, A., Bliven, S.E., et al.: BioJava: an open-source framework for bioinformatics in 2012. Bioinformatics **28**, 2693–2695 (2012)
13. Schatz, M.C.: CloudBurst: highly sensitive read mapping with MapReduce. Bioinformatics **25**(11), 1363–1369 (2009)
14. Shindyalov, I., Bourne, P.: Protein structure alignment by incremental combinatorial extension (CE) of the optimal path. Protein Eng. **11**(9), 739–747 (1998)

15. Sosinsky, B.: Cloud Computing Bible, 1st edn. Wiley, New York (2011)
16. Microsoft Azure Cloud Services Specification: Virtual Machine and Cloud Service Sizes for Microsoft Azure. http://msdn.microsoft.com/en-us/library/windowsazure/dn197896.aspx (2013). Accessed 25 Nov 2013
17. Microsoft Azure Specification. http://msdn.microsoft.com/en-us/library/windowsazure/dd163896.aspx (2013). Accessed 25 Nov 2013
18. Ye, Y., Godzik, A.: Flexible structure alignment by chaining aligned fragment pairs allowing twists. Bioinformatics **19**(2), 246–255 (2003)
19. Zou, Q., Li, X.B., Jiang, W.R., Lin, Z.Y., Li, G.L., Chen, K.: Survey of MapReduce frame operation in bioinformatics. Brief Bioinform. 1–11 (2013)

Chapter 5
General Discussion and Concluding Remarks

At its essence, the field of bioinformatics is about comparisons.

Jonathan Pevsner, 2009 [1]

Abstract In this chapter, I will try to summarize what we talked about through all four chapters. The chapter summarizes various types of parallelisms that were used while comparing proteins based on different features of their structures. I also give some advantages and drawbacks of the presented high-performance computational solutions for protein comparison, alignment, matching, and similarity searching.

5.1 General Discussion

Numerous solutions for protein similarity searching prove that it is one of the important tasks performed in the domain of protein bioinformatics. The process can be carried out due to various reasons, like protein identification, protein function identification, phylogeny reconstruction, and others. It can be also supportive for protein structure modeling. The computational complexity of a plethora of methods that were developed for protein similarity searching, protein alignment, and protein structure matching implies the necessity of using advanced techniques and computational architectures to complete these tasks in a reasonable time. In this book, we could see some of the techniques and architectures that benefit from the recent achievements in the field of computing and parallelism.

It is worth noting that techniques and methods presented in the successive chapters of this book are based on various types of parallelism. The search engine for the PSS-SQL language uses multithreading during the calculation of the similarity matrix. The matrix is divided into areas, these areas are assigned to multiple threads, and the calculation of the whole matrix is parallelized on multicore CPU. Multiple threads calculate one similarity matrix. The calculation of areas is dependent on the calculation of other areas in the matrix, which requires a synchronization of threads

D. Mrozek, *High-Performance Computational Solutions in Protein Bioinformatics*,
SpringerBriefs in Computer Science, DOI: 10.1007/978-3-319-06971-5_5,
© The Author(s) 2014

and slows down the calculation process. However, upon omitting disadvantages, this solution is faster than a single-core implementation and portable to any PC computer, since it adapts to the number of cores possessed by the user.

On the other hand, the GPU-CASSERT makes use of many-core GPU devices and multiple threads for finding structural similarities between the given protein and proteins from the database. In this solution, regardless of the phase of the matching algorithm, the similarity matrix for a pair of compared protein structures is calculated by a single thread. Multiple threads, working in parallel, calculate multiple similarity matrices for multiple database proteins. Threads are completely independent. This allows to avoid costly synchronizations, increases performance of the solution, but this also requires a lot of macromolecular data to be transferred to the GPU device at the same time, which implies the use of a low-bandwidth, off-chip global memory. Similarly to the PSS-SQL search engine, the GPU-CASSERT also divides each similarity matrix into areas of the fixed size. However, it does so because the number of registers per thread is limited and GPU-CASSERT tries to minimize the number of read/write transactions to the memory structures (especially global memory) of the GPU device. GPU-CASSERT requires dedicated GPU devices with the CUDA compute capability, which is not available on every workstation. However, it provides much better efficiency than its CPU-based counterpart.

Finally, Cloud4Psi utilizes many instances of virtual machines that serve as compute units. Each Searcher role, as a compute unit, works independently of every other Searcher role. While looking for protein similarities, Searchers execute in parallel the same logic of protein structure comparison, alignment and superposition for different 3D protein structures from the data repository. Actually, the Cloud4Psi is a representative of the parametric sweep application and the entire process is embarrassingly parallel or delightfully parallel. The Manager role passes different initial parameters to the Searchers working on each compute node through the queueing system. This lets each Searcher apply the same logic to different macromolecular data. To avoid excessive exchange of messages, each Searcher performs calculations for a group of proteins from the repository, one by one. Many Searchers work concurrently, processing separate groups of protein structures. The number of Searchers depends on the amount of available cloud resources. At the moment, we are working on the more efficient version of the Cloud4Psi that will join two types of parallelism, i.e., independent computations for a group of proteins from the repository (parametric sweep) and multithreaded, independent calculations for proteins inside each group performed on many-core CPUs. In such a way, the Cloud4Psi could be scaled out by adding more compute units and efficiently scaled up by using compute units of higher compute capabilities. Of these three solutions, Cloud4Psi requires the computational resources that are not available to everyone. However, it gives the possibility of unlimited scaling out.

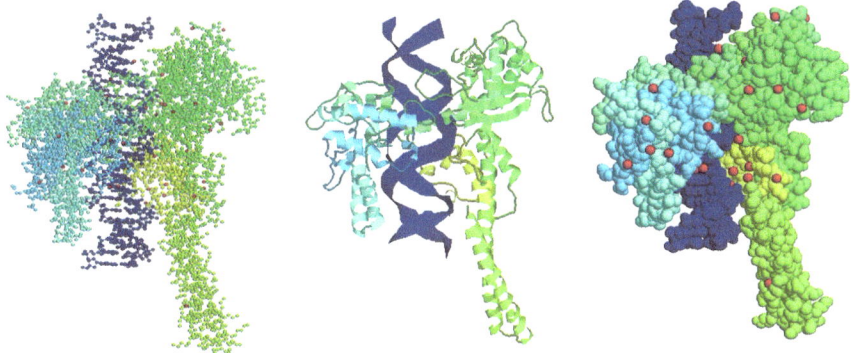

Fig. 5.1 Beautiful structure of the human topoisomerase I/DNA complex [PDB ID: 1A36] [3] responsible for relaxing and untangling DNA strands in the nucleus. Visualized using RasMol [2]—representation modes: (*left*) atomic, (*middle*) ribbons (secondary structures), (*right*) spacefill

5.2 Concluding Remarks

Beautiful structures of proteins, like the one presented in Fig. 5.1, are definitely worth creating efficient methods for their exploration and analysis, with the aim of mining the knowledge that will improve human life in further perspective. While writing this book, I tried to pass through various representation levels of protein structures and show methods suitable for the particular level. In the successive chapters of the book, I described methods that were developed either by myself or as a part of projects that I was involved in. Certainly, there are other solutions for the presented problems, which I referenced in particular chapters, but I hope that the solutions presented in the book will turn out to be interesting and helpful for scientists, researchers, and software developers working in the field of protein bioinformatics.

Supplementary materials are available from home pages of a particular project:

- PSS-SQL project home page:
  http://zti.polsl.pl/dmrozek/science/pss-sql.htm
- GPU-CASSERT project home page:
  http://zti.polsl.pl/dmrozek/science/gpucassert/cassert.htm
- Cloud4Psi project home page:
  http://zti.polsl.pl/dmrozek/science/cloud4psi.htm

References

1. Pevsner, J.: Bioinformatics and Functional Genomics, 2nd edn. Wiley-Blackwell, Boston (2009)
2. Sayle, R.: RasMol, Molecular graphics visualization tool. Biomolecular Structures Group, Glaxo Welcome Research & Development, Stevenage. http://www.umass.edu/microbio/rasmol/ (1998). Accessed 5 Feb 2013
3. Stewart, L., Redinbo, M.R., Qiu, X., Hol, W.G., Champoux, J.J.: A model for the mechanism of human topoisomerase I. Science **279**(5356), 1534–1541 (1998)

Index

D. Mrozek, *High-Performance Computational Solutions in Protein Bioinformatics*,
SpringerBriefs in Computer Science, DOI: 10.1007/978-3-319-06971-5,
© The Author(s) 2014